W0016666

To My Dear Friend
Professor Shu-Park Chan
With my Best Wishes,
Nam Ling
Aug, 2000

樹柏藏書

二〇〇〇年秋書於加州矽谷

國際科技大學

作者林楠博士贈閱此冊

[印章]

# Specification and Verification of Systolic Arrays

# Specification and Verification of Systolic Arrays

**Nam Ling**
*Santa Clara University*

**Magdy A Bayoumi**
*University of Southwestern Louisiana*

**World Scientific**
*Singapore • New Jersey • London • Hong Kong*

*Published by*

World Scientific Publishing Co. Pte. Ltd.

P O Box 128, Farrer Road, Singapore 912805

*USA office:* Suite 1B, 1060 Main Street, River Edge, NJ 07661

*UK office:* 57 Shelton Street, Covent Garden, London WC2H 9HE

**British Library Cataloguing-in-Publication Data**
A catalogue record for this book is available from the British Library.

**SPECIFICATION AND VERIFICATION OF SYSTOLIC ARRAYS**

Copyright © 1999 by World Scientific Publishing Co. Pte. Ltd.

*All rights reserved. This book, or parts thereof, may not be reproduced in any form or by any means, electronic or mechanical, including photocopying, recording or any information storage and retrieval system now known or to be invented, without written permission from the Publisher.*

For photocopying of material in this volume, please pay a copying fee through the Copyright Clearance Center, Inc., 222 Rosewood Drive, Danvers, MA 01923, USA. In this case permission to photocopy is not required from the publisher.

ISBN 981-02-3867-3

Printed in Singapore.

To my parents, my wife, and my daughter

(Nam Ling)


To my parents

(Magdy A. Bayoumi)

# TABLE OF CONTENTS

# PREFACE

In the past decade, the demand for high-speed computation in many digital and image processing applications has increased dramatically. In many applications, systolic arrays proved to be effective answers to such challenges. Systolic arrays offer both parallelism and pipelining required for high-speed computation. A systolic array is a network of simple processors, which rhythmically compute and pass data through the system. The advantages of using systolic arrays for high-speed computation can be seen at three different levels: algorithmic, architectural, and technological.

On the technological level, the availability of low-cost, high-density, high-speed very large scale integration (VLSI) devices and efficient computer-aided design (CAD) facilities result in many revolutionary changes. VLSI technology is especially suitable for designs which are *regular*, *repeatable*, and with highly *localized* communications. Such structures reduce design time, improve interconnect delay, and ease synchronization.

On the algorithmic level, most signal and image processing algorithms are dominated by transform techniques, convolution/correlation filtering, and some key linear algebraic methods. These algorithms require enormous throughput rates and huge amount of data and memories. Interestingly, most of these algorithms are *regular*, *recursive*, and *localized*, which are extremely suitable for array processor designs, and in particular, systolic array designs.

Finally, on the architectural level, systolic arrays belong to a class of highly parallel application-specific array architectures that exploit the nature of many algorithms to achieve a high degree of pipelining and parallelism for highly intensive computation. Moreover, systolic arrays balance computation and input/output (I/O) bandwidths, which further improve system performance.

In spite of the large amount of research work, not many systolic array books have been published. Most of these books are collections of papers, dealing with a wider range of topics besides systolic arrays. We believe that a comprehensive book dedicated to systematic formal (mathematical) techniques in systolic array design specification and verification is important.

Formal (mathematical) techniques are important for precise and complete description of designs as well as for ensuring design correctness. Although simulations and experimental methods have been used widely as techniques for checking hardware and architectural designs, they are usually very time-consuming and do not guarantee the conformance of designs to their algorithmic level specifications. With the increasing complexity of hardware and the importance of design time, design correctness has become increasingly significant, as errors in a design may result in strenuous debugging processes, or even in the repetition of a costly manufacturing process. *Formal* description and verification of architectures is therefore vital as this is the only way to ensure unambiguous and complete communications among designers and implementers, as well as the full conformance of designs to specifications. The increase in formal techniques being adopted in CAD tools and in hardware design industry in the past decade attests to their importance.

In the case of systolic arrays, although many techniques have been developed to produce correct-by-construction arrays from given algorithms, these techniques can only be applied to a limited class of algorithms. In fact, many systolic arrays are designed by ad hoc or systematic, but not necessarily formal, techniques. Formal methods are therefore necessary to serve as tools to guarantee the correctness of systolic designs. With an efficient formal specification and verification tool for systolic architecture, ensuring the correctness of systolic designs can be made possible and design and debugging time can be significantly reduced. This meets the demand and challenge for ensuring the correctness of complex arrays designed.

In our book, we first describe systolic arrays and their design issues. A survey and review is presented on different formal methods proposed in the past for digital circuits/modules, as well as those for systolic architectures, together with their strengths and weaknesses. A novel formalism, named *Systolic Temporal Arithmetic (STA)*, developed by the authors, is then introduced. Useful STA axioms, rules, and theorems are also presented and proved. The formalism exploits systolic properties to provide constructs and verification techniques for effective and efficient specification and verification of systolic array designs. A framework for formal specification and verification of systolic arrays is then produced. Three verification techniques, output derivation and comparison, mathematical induction, and solving STA difference equations, are then described. Several application examples are provided to illustrate our framework and verification techniques. A Prolog-based verifier, named *Verifier for STA (VSTA)*, developed with the help of Dr. Timothy Shih, is presented to provide a CAD facility for semi-automated interactive design verification. Four inductive techniques adopted by our verifier are also defined. Finally, we provide an example showing how a complex systolic array designed for LU decomposition can be specified and verified by our technique. A sample output from *VSTA* and a brief description of its user interface are given in the appendix.

The book can serve as a reference book for researchers and designers. It can also be used as a supplementary textbook for computer engineering, computer science, or electrical engineering graduate courses in systolic arrays or other specialized courses in parallel architectures or formal hardware verification.

# ACKNOWLEDGEMENTS

We would like to express our special gratitude to Dr. Timothy Shih, currently Associate Professor at Tamkang University, Taiwan, for his help in developing the verifier, while he was Dr. Ling's research assistant at Santa Clara University. We also thank Dr. Ruth Davis, Dr. Shih's Ph.D. supervisor while he was at Santa Clara, and Dr. Dan Lewis, the Chair of the Computer Engineering Department. Thanks also go to Drs. Subrata Dasgupta and N. A. Ramakrishna, for constructive discussions and help during the earlier stage of the project at the University of Southwestern Louisiana, U.S.A. We would like to acknowledge the support from two U.S. National Science Foundation (NSF) grants (MIP-8809811 to the University of Southwestern Louisiana for Dr. Bayoumi, and MIP-9010385 to Santa Clara University for Dr. Ling) as well as the Arthur Vining Davis Junior Faculty fellowship, given to Dr. Ling, in support of this research at Santa Clara University. We would also like to acknowledge the support from students and faculty from the School of Engineering and the Computer Engineering Department (Santa Clara University, U.S.A), as well as the Center for Advanced Computer Studies (University of Southwestern Louisiana, U.S.A). Dr. Ling would also like to thank the Center for Signal Processing and the School of EEE (Nanyang Technological University, Singapore), where he spent his sabbatical leave during part of 1998. The earlier stage of the research work was performed at the University of Southwestern Louisiana and the later stage of the work was carried out at Santa Clara University. Part of the writing was carried out in Singapore.

Our deepest gratitude goes to God for His love and our families for their support, care, help, and love. Dr. Ling would like to thank his wife Mei-Yan, his daughter Grace, and his parents Yu Chich and Siew Chee for their support and love. Dr. Bayoumi would like to thank his wife Seham and his kids Aiman, Walid, and Amanda for keeping up with his crazy schedule. We would also like to thank World Scientific for publishing our materials.

NAM LING
Santa Clara University
Santa Clara, California
U.S.A.

MAGDY A. BAYOUMI
University of Southwestern Louisiana
Lafayette, Louisiana
U.S.A.

# Chapter 1

# INTRODUCTION

## 1.1   INTRODUCTION

The availability of low-cost, high-density, high-speed very large scale integration (VLSI) devices [Mead80] and efficient computer-aided design (CAD) facilities presages a major breakthrough in the design and application of massively parallel processors. In particular, VLSI microelectronic technology has inspired many innovative designs in array processor architectures [KungS88]. One major class of such architectures is systolic arrays [KungH78, KungH82], which serve as efficient means in realizing many algorithms in modern signal and image processing applications.

According to Kung and Leiserson [KungH78], "A systolic system is a network of processing elements (PEs or cells) which rhythmically compute and pass data through the system." Each systolic array is usually dedicated to a particular application algorithm. Systolic arrays have the important properties of modularity, regularity, local interconnection, a high degree of pipelining, and highly synchronized multiprocessing. Systolic array design differs from the conventional von Neumann computer in its highly regular and pipelined computation [KungS88]. Once a data item is brought out from the memory it can be used effectively at each cell it passes while being "pumped" from cell to cell along the array. This is especially appealing for a wide class of compute-bound computations, where multiple operations are performed on each data item in a repetitive manner (in a computation, if the total number of operations is larger than the total number of inputs/outputs, then the computation is compute-bound, otherwise it is I/O-bound). Such processing avoids the classic memory access bottleneck problem commonly incurred in von Neumann machines and speeds up many compute-bound computations. Moreover, systolic array design is generally simple due to its highly synchronized nature; it is well suited for VLSI, due to its modularity, regularity, repeatability, and locality nature. Features like these simplify designs and resolve many propagation delay problems in VLSI.

## 1.2   DEFINITION OF SYSTOLIC ARRAYS

There are a number of "definitions" of systolic arrays [KungH78, Ull84, Rao85, KungS88]. To give a coherent definition for further discussion, we adopt the following definition and properties:

**Definition** [Rao85]: A systolic array is a network of processors in which the processors can be placed at the grid points of a finite (multi-dimensional) lattice so that

- Topologically: If there is directed link *from* the processor at location *I* *to* the processor at location *(I+D)* for some *I* and for some constant *D*, then there is such a link for every *I* within the lattice, and
- Computationally: If a processor receives a value on an input link at time *t*, then it receives a value at time *(t+∆)* on the same input link and places an output value at time *(t+∆)* on the corresponding output link, where ∆ is the fundamental time-period that is independent of the size of the network, the orientation of the link, or the location of the processor.

Informally speaking, a systolic array is a computing network possessing the following features [KungS88]:

- *Synchrony*: A systolic array is controlled and synchronized by a global clock with fixed length clock cycles. Data are rhythmically computed (timed by a clock) and passed through the systolic array network. The clock signal serves two purposes; as a sequence reference and also as a time reference. As a sequence reference, the clock transitions serve the purpose of defining successive instants at which system state changes may occur. As a time reference, the period between clock transitions accounts for wiring and element delays in paths from the output to input of clock elements (cells).
- *Modularity and Regularity*: A systolic array consists of modular processing units with homogeneous interconnections and the computing network can be extended indefinitely.
- *Spatial locality and temporal locality*: A systolic array manifests a locally communicative interconnection structure, i.e., spatial locality. Each cell or processing element (PE) only communicates with its immediate neighboring cells. There is at least one unit-time delay allotted so that signal transactions from one cell to the next can be completed, i.e., temporal locality.
- *Pipelinability*: A systolic array exhibits a linear rate pipelinability to speed up processing rate, i.e., it should achieve an $O(n)$ speed up, in terms of processing rate, where *n* is the number of processing elements.

In addition to the above properties, many systolic arrays also exhibit the following features:

- *Repeatability*: In most systolic arrays, the entire network is usually the repetition of one single type of cell (PE) and interconnection. Even in the cases of other systolic arrays, there are usually at most two or three different types of PEs involved in a network.
- *Parallel Processing*: In the case of two-dimensional (2-D) or multidimensional systolic arrays [Ling88], higher parallelism can be achieved by processing the rows/columns/planes concurrently.

## 1.3    SYSTOLIC ARRAY ABSTRACTION LEVELS

A systolic array can be abstracted at different architectural levels [KungS88] as shown in Figure 1.1:

1. The *Algorithm level* (level 3) defines the algorithm/mathematical expression to be realized.
2. The *Array level* (level 2) defines the interconnection between different cells (PEs) and their functional capabilities.
3. The *PE or Cell level* (level 1) defines the hardware modules (e.g. adders, multipliers, registers) for the PEs and their interfaces.
4. The *Module and Circuit level* (level 0) defines the internal circuitry of individual modules.

A further *Geometric or Layout level* can be defined which specifies the photolithography mask features of the circuitry for VLSI fabrication. At each of the abstraction levels the architecture is hierarchically specified to decrease the complexity of the description. Only at the *array level* that an architecture depicts the properties of a systolic array as spelled out in Section 1.2.

From the implementation/specification point of view, the implementation at level $i$ becomes the specification of level $i$-1 and the specification at level $i$ is the implementation of level $i$+1. The design of a particular architectural level $i$ involves the realization of the specification of level $i$+1 with the architecture of level $i$ and proving the correctness of the implementation at level $i$ with respect to the specification at level $i$+1.

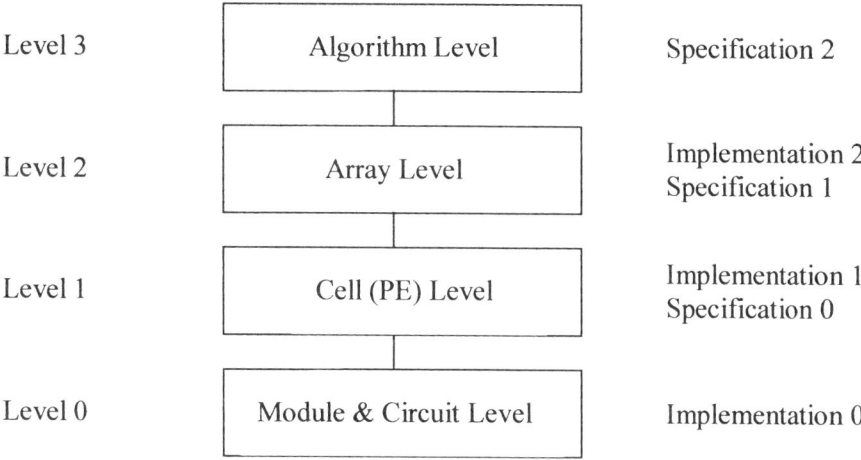

Figure 1.1 Different architectural levels of a systolic array

## 1.4   THE DESIGN OF SYSTOLIC ARRAYS

This book concerns only with the design of the *array level* architecture, given the algorithm or mathematical expression. From the definition and description of Sections 1.2 and 1.3, we can view a systolic array array level architecture $A$ as a 6-tuple:

$$A = < C, COMPT, CFUNC, INC, IN, OUT >$$

where:

$C$ is the global clock which synchronizes and controls the array in the way defined by the systolic temporal frame $T_s$ (discussed in Chapter 3),

$COMPT$ is a set of components (e.g. PEs) used to build the array,

$CFUNC$ is a set of component functions,

$INC$ defines the regular local interconnection pattern for the components, including the locations of array inputs/outputs,

$IN$ is a set of array data inputs, including the timings and the locations of these inputs, and

$OUT$ is a set of array result outputs, their timings and locations.

Detailed discussions of these entities are provided in Chapters 3 and 4.

For the design of array level architecture, we adopt the most widely accepted design paradigm [Dasg88a], which consists of three main phases: analysis, synthesis, and evaluation [Jone63]. We suggest an additional phase, termed description, to be included in our design process. Hence our design process consists of four main phases: *analysis, synthesis, description,* and *evaluation*. These correspond to *problem analysis, synthesis (mapping), specification,* and *verification* in systolic array design, as depicted in Figure 1.2. Any decision or problem occurred in any stage may require the designer to return to an earlier phase for appropriate modification or clarification. These four phases are discussed below:

1. *Problem Analysis*: This phase requires the look into the problem of concern in the area of signal processing, image processing, or scientific computation. The factors and the parameters affecting the problem must be studied. The result of this phase is a mathematical expression or an algorithm for solving the problem (the algorithm level architecture).

2. *Synthesis (Mapping)*: Given suitable mathematical expression or algorithm for solving the problem, synthesis (mapping) is a process of converting this into a systolic array architecture. In other word, each entity in the 6-tuple of $A$ must be defined at the end of this phase.

3. *Specification*: This is a description of the systolic array architecture synthesized from the above phase. A good specification should provide *precise* and *complete* description of the target architecture. To guarantee a *precise* specification of the architecture, formal (mathematical) notation is necessary to prevent any ambiguous interpretation. To have a *complete* specification of the architecture, no essential information necessary for the construction of the array that will correctly realize the algorithm level mathematical specification should be left out. In other words, one should at least specify:

   (a) How the array is to be "constructed": This is an abstract description of the physical design, which is intended to serve as a basis for realization. That is, one should specify the type of components (PEs), and the number of each that should be used to build the array, as well as the way they should be connected (defining the entities $COMPT$, $CFUNC$, and $INC$ of $A$).

   (b) How the array is to be "used": That is, when, where, and what data/results should be input/output to/from the array (defining the entities $IN$ and $OUT$ of $A$).

   The clock $C$, defined by systolic temporal frame $T_s$ (Section 3.2), is standard and implicit for all arrays and need not be explicitly specified. Other features such as word length and clock duration can also be specified at the array level if they are vital for the correct realization of the algorithm. Precise and complete specification is vital for communications among designers and implementers, so that precise and complete description can be given to the implementers for implementing the array, and such information can be given to the designer for verifying the correctness of the array, or redesign, if necessary.

4. *Verification*: This is a process for checking the conformance of the array architecture with respect to its upper level algorithm (mathematical) specification (design correctness). Design correctness has become increasingly significant as errors in design may result in a strenuous debugging process, or even in the repetition of a costly manufacturing process. Although simulations and experimental methods have been used widely as techniques for checking hardware and architectural designs, they do not guarantee the conformance of designs to upper level specifications. To ensure that an array level design correctly realizes its upper level algorithm specification, formal verification techniques are necessary. A good formal verification technique should be *sound* so that the correctness of the results can be guaranteed. Therefore, all the rules, theorems, and derivation steps used in the verification must be sound or mathematically proved.

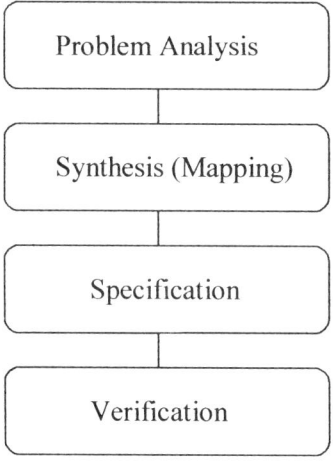

Figure 1.2 Systolic array design phases

## 1.5   THE OBJECTIVE OF THE BOOK

Research advances in signal processing, image processing, algorithm analysis/synthesis, and many scientific computation areas have resulted in many well developed techniques in producing systolic algorithms for solving suitable problems in these areas. Many solid mapping techniques have also been developed in the past decade for synthesizing suitable systolic algorithms to systolic architectures. However, well-developed techniques for formal specification and using it for formal verification of systolic architectures are still rare and have some limitations (a review is provided in Chapter 2). The main objective of the book is thus concentrated on developing a solid formalism and verification technique for formal specification and verification of systolic architectures so that users can combine our technique with an already well developed mapping and problem analysis technique for complete design of systolic arrays. We call our formalism *Systolic Temporal Arithmetic* (STA) [Ling89b, Ling90] due to the fact that it is based on describing arithmetic operations in a systolic dynamic environment. Hence, given a systolic algorithm, a systolic array designer can apply any suitable mapping technique (even an ad hoc technique) to derive a suitable systolic architecture. One can then specify the architecture using STA and verify its correctness with respect to the algorithm by our STA verification techniques. Figure 1.3 depicts how STA can be used as part of a design tool.

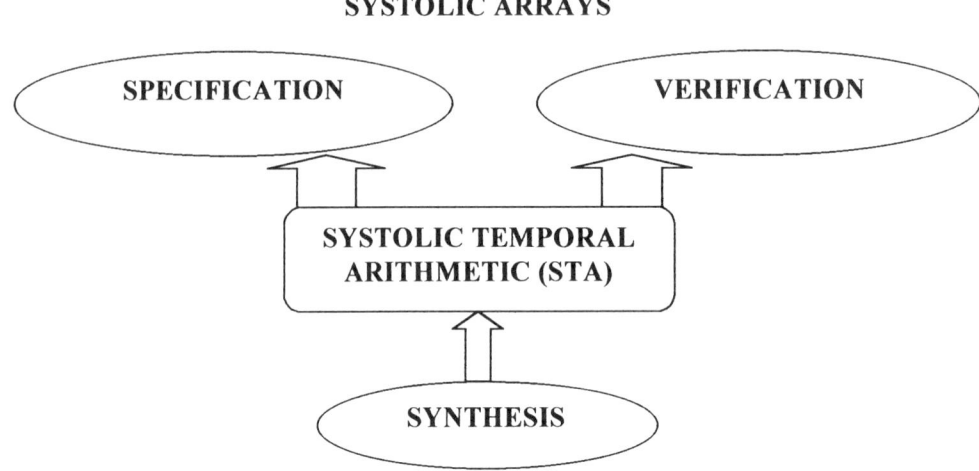

Figure 1.3 Systolic Temporal Arithmetic for systolic array design

The objectives of STA are, firstly, to serve as a specification tool so that, given a systolic array design at the array level, one can specify the design such that:

1. The description of the design is complete, precise, and suitable for unambiguous communications among designers and implementers.
2. The specification is simple and lengthy descriptions can be avoided.
3. The notation and the underlying semantics are coherent and can be used for verifying the conformance of design to the algorithm specification in a formal way.
4. The notation and the underlying semantics can preferably also be used for other purposes such as simulations, fault diagnoses, and test generations.

Secondly, given a systolic array design at the array level, one can formally verify the correctness of the design with respect to its algorithm level specification using this tool, satisfying the following aims:

5. The technique is sound.
6. The technique exploits the unique features of systolic arrays to produce notation and verification procedures that are simple, elegant, and efficient.
7. The constructs and operators involved can preferably be unified with one useful lower level formalism to form a coherent multilevel reasoning system.
8. The notation used deals with inherent parallelism and pipelining in an array and does not impose any restriction on array design flexibility.
9. The procedure can be automated and can be incorporated into CAD packages.

## 1.6   AN OVERVIEW OF THE BOOK

Chapter 1 presents a brief description of systolic arrays and their design issues. The definition and the properties of systolic arrays are given. Different architectural abstraction levels of systolic architectures are mentioned. Four main phases of design: problem analysis, synthesis (mapping), specification, and verification are discussed. A general outline of our research objectives is introduced.

In Chapter 2, a review on published work on formal specification and verification of digital circuits/modules is provided, followed by a review on existing techniques on formal specification and verification of systolic architectures at the array level. The limitations of existing techniques are discussed and the motivation and challenge for developing a suitable formalism, as well as the research direction, are given. Abstraction mechanisms adopted for the development of our formalism and how systolic features are exploited are also discussed.

Chapter 3 describes our new formalism, termed *Systolic Temporal Arithmetic* (STA), for formal specification and verification of systolic arrays. The model of time for this formalism is defined. Syntax and semantics of STA are presented. Axioms, rules, and theorems developed are discussed. Proofs are provided whenever it is necessary. Useful constructs are also developed to exploit systolic features to simplify specifications and verifications.

Chapter 4 contains a presentation of the framework developed for formal specification and verification of systolic arrays. Several major techniques developed for efficient systolic array verification: the output derivation and comparison method, the mathematical induction method, and the method by solving STA difference equations, are presented. The induction method is further expanded into four different inductive techniques to suit different array topologies and data flow.

Chapter 5 applies STA to specify and verify several systolic array architectures, which are common in digital signal processing. These include a 1-D systolic array for matrix-vector multiplication, a 2-D systolic array for matrix-matrix multiplication, and a bi-directional systolic array for 1-D convolution.

In Chapter 6, a Prolog-based semi-automated verifier developed, called VSTA, is described. Our verifier uses induction, backward chaining, and rewriting, to perform a proof of the goal.

In Chapter 7, our technique is applied to verify the correctness of a complicated systolic array designed for LU decomposition.

Chapter 8 summarizes our results. VSTA user interface as well as a sample session of our verifier specifying and proving the correctness of a 2-D array for matrix multiplication is given in Appendix A.

# SPECIFICATION AND VERIFICATION OF SYSTOLIC ARRAYS: DEFINITIONS AND RELATED WORK

## 2.1    INTRODUCTION

As a result of the increasing demand in high-speed computation in the areas of digital signal processing, image processing, and scientific computation, and the advent of multidimensional systolic arrays for performance enhancement [Ling88, Ling89c], systolic arrays have become more complex in terms of cell structure, interconnection topology, and data flow. As explained in Chapter 1, complete, precise and formal description of architectures has become vital for unambiguous and complete communications among designers and implementers, as well as for formal verification of correctness. Design correctness has also become increasingly significant as errors in design may result in a strenuous debugging process, or even in the repetition of a costly manufacturing process. Although simulations and experimental methods have been used widely as techniques for checking hardware and architectural designs, they do not guarantee the conformance of designs to upper level specifications. All these point to the need for formal specification and verification: a mathematical (instead of an experimental) approach for describing and reasoning of architectures. A formalism for such purpose should aim to be simple, complete, precise, coherent, automatable, efficient, and reliable. We now provide definitions for formal specification and verification of systolic array array level architecture for our discussions:

**Definition:** *Formal specification* of a systolic array array level architecture is a formal (mathematical) description of the architecture. Such a description must be precise and complete. To guarantee preciseness, formal (mathematical) notation should be used for the description to prevent any ambiguous interpretation. To have a complete description, no essential information necessary for the construction of the array that will correctly realize the specified algorithm should be left out: both physical design description (describing the type of components (PEs) used, the number of each, and the way they should be connected) and array functional description (describing when, where, and what data/results should be input/output to/from the array) must be included.

**Definition:** *Formal verification* of a systolic array array level architecture is a formal (mathematical) process for checking the conformance of the array level architecture design to its upper (algorithmic or mathematical) level specification (design correctness). Formal verification can take a formal specification of the architecture as input, to produce a result to indicate whether the array level design correctly realizes the algorithm level mathematical specification. To guarantee the correctness of the result, all rules, theorems, and derivation steps, etc., used in the verification must be sound or mathematically proved.

In our context, we restrict our discussions to the verification of functional correctness through formal techniques, which is given in the above definition. A systolic array can be abstracted at various architectural levels as discussed in Section 1.3 and shown in Figure 1.1. The implementation at level $i$ becomes the specification of level $i$-1 and the specification at level $i$ is the implementation of level $i$+1. Any piece of hardware is functionally correct if we can prove that its implementation realizes the upper level specification.

## 2.2    A REVIEW OF RELATED WORK

### 2.2.1  Related Work in the Specification and Verification of Digital Circuits and Modules

Many techniques have been developed for the specification and verification of digital circuits and modules [Yoe90, Gupt92]. Techniques adopting first order predicates include Wagner's proof checker [Wag77], Wojcik's Aura system [Woj84], and Uehara et al.'s DDL Verifier [Ueha83] using backward reasoning. Eveking [Evek85] considers how hardware description languages can be applied to abstract hardware specification for higher-level descriptions. Melham [MelT88, MelT93] argues that techniques for proving the correctness of hardware designs must use abstraction mechanisms for relating formal descriptions at different levels of detail. Higher-order predicates have also been applied for circuit verifications [MelT93]; examples include Hanna and Daeche's Veritas [Hann86] and Gordon's HOL [Gord86]. Besides these, many authors create their own formal systems: for example, Milne's Circal [Mil83], Gordon's LCF-LSM system [Gord83], and Barrow's Prolog-based Verify Program [Barr84]. Software verification techniques have also been applied to hardware verifications; Shostak's graphical model of hardware [Shos83], where Floyd's induction assertion method is applied, is an excellent example.

Later trend in hardware specification and verification involves the use of temporal logic [Res71, Mann81, Turn84], which is a generalization of predicate logic to encompass the temporal domain for effective description of dynamic environments. Temporal logic substantially advances traditional logic because it can capture time and dynamic behaviors – essential features in hardware descriptions - with concise and clear notation [Camu88]. It avoids the introduction of explicit time functions and time variables. Bochmann [Boch82] presents one of the earlier work in using linear and discrete time temporal logic for hardware description and reasoning; Fujita et al. [Fuji83] demonstrate an approach to formal verification based on linear temporal logic and Prolog, and propose a total CAD system for hardware design based on temporal logic, DDL, and gate-level description; Malachi and Owicki [Mala81] utilize temporal logic to model self-timed digital systems; Browne et al. [Brow86] describe an automatic verification system using temporal logic for sequential circuit verification; Moszkowski [Mosz83, Mosz85, Halp83] introduces a general-purpose formalism, called ITL (interval temporal logic), for reasoning hardware behavior; he also develops Tempura, a logic programming language for programming concepts of ITL.

Besides the above-mentioned techniques, architectural description languages (ADLs) have also been introduced by many authors for specifying computer hardware and modules. Proof rules are also being developed for some ADLs for the purpose of verification [Broc92]. Examples of ADLs include Bell and Newell's ISP [Bell71], Duley and Dietmeyer's DDL [Dule68], Shahdad et al.'s VHDL [Shah85], and Dasgupta et al.'s axiomatic specification and S*M [Wilse87, Dasg87, Dasg88b]. Occam is also considered [Wilso83] for architectural or module descriptions.

## 2.2.2  Related Work in the Specification and Verification of Systolic Arrays

Although many techniques have already been developed for specification and verification of circuit and module level hardware, work on specification with functional verification of systolic arrays at the array architectural level are still rare. Some significant techniques have been developed in the past. Many techniques use one or more of the three major approaches:

1.  Using mathematical equations and sequences.
2.  Using the concept of computational wavefront.
3.  Using languages and software techniques.

One of the most significant contributions in this area is the mathematical model of systolic networks due to Melhem and Rheinboldt [Melh84]. Data items are represented by data sequences and cell computations are modeled by a system of difference equations involving operations on these data sequences. The input/output of the network are obtained by solving this system of difference equations. The correctness of the network is verified by the correct input/output description obtained.

Kuo, Levy, and Musicus [Kuo84] formulate space-time-data equations to describe the motion of data elements and waves in an array; using this approach, the correctness of a systolic architecture implementing a desired algorithm can be proved. Jover, Kailath, Lev-Ari, and Rao [Jove86] use equivalence transformations to convert an algorithm to an array. Verification is performed by reversing the transformation path to obtain an algorithm equivalent to the original one.

Hennessey [Henn86] describes systolic arrays by a language called CCS (Calculus for Communicating Systems, due to Milner [Miln80, Miln83]), which is very similar to INMOS's Occam [Wilso83]. He applies syntactic transformations for verification. Such a set of transformations can be used to transform specification descriptions to implementation descriptions, or vice versa. Chen and Mead [Chen82, Chen83] embodies the idea that each processing element implements a set of functions and that the behavior of a whole network is a set of recursive equations obtained by analyzing the interconnection of the processing elements. Inductive techniques for program verifications are used to show the correctness of systolic algorithms. Rajopadhye and Panangaden [Rajo85] express processing element as a function on a stream of data; they also use program verification techniques to verify the correctness of systolic arrays, each of which is a connection of these processing elements.

Other related results are due to Kung and Leiserson [KungH78], Lev-Ari [Lev83], Kung and Lin [KungH83], Tiden [Tid84], Probst and Li [Prob88], Jover and Kailath's Lines of Computation

(LOCs) concept [Jove84], Ossefort and Foster's program verification techniques [Osse82, Fort81], Purushothaman and Subrahmanyam's mechanical certification and the use of Boyer-Moore theorem prover [Puru89], and a few others.

Provided above are a review on the techniques for systolic array array level verification and specification notations developed which can be used for verifications. However, many authors have applied formal notation to specifying systolic arrays and used them for synthesis purposes. Examples include [Mold83], [Mold84], [Quin84], [Capp83], [Capp84], [Rao85], [KungS85], and many others. Some mapping procedures will result in guaranteed correct arrays being synthesized for certain classes of algorithms and the space-time representation of the arrays are described using well-defined notation.

## 2.3  MOTIVATION AND CHALLENGE

A formalism for the purpose of specifying and verifying a systolic array at the array architectural level requires a complete, precise, and coherent definition of the underlying semantics [Camu88]. The formalism should also be simple, automatable, and exploits the nature of the target architectures to provide simple, elegant, and efficient notation and verification procedures. Moreover, it should preferably use similar notation and semantics with a lower level formalism so that they can be unified to form a sufficiently coherent multilevel reasoning system for various levels of hardware. This can help to provide unambiguous communications among designers and implementers of various levels. It can also help to facilitate coherent multilevel formal verifications, which is important since the implementation correctness of an array necessitates the verifying of its building block circuits. Moreover, it can help to reduce the need to be familiarized with two or more substantially different formalisms. (One might think of developing a formalism that can be utilized for many levels of hardware, but such a formalism is usually large, complex, and does not exploit the properties that belonged to a particular architectural level). Most existing work on array level specification and verification have the following common limitations:

1.  Many existing tools do not fully exploit unique systolic array features such as synchrony, regularity, repeatability, modularity, pipelinability, parallel processing ability, as well as spatial and temporal locality to provide constructs to simplify specifications and to provide short-cuts in verifications.
2.  Many tools (especially those using transformation graphs, matrices, and mathematical equations) and their notations that are used for array level descriptions are not suitable for lower (circuit) level details and it is difficult to unify them with other lower level formalisms to form a coherent multilevel specification and verification system.
3.  Some techniques also impose inflexibilities in array designs.

The challenge of the work presented in this book is to produce a formalism and related methods to overcome these limitations to achieve simple, complete, and precise specification for systolic designs as well as efficient and reliable verification scheme for these designs. A secondary challenge is to automate our techniques and produce a verifier.

## 2.4    RESEARCH DIRECTION

The formalism, as well as the specification and verification methods (based on this formalism) we developed, aim on the objectives stated in Section 1.5 and on overcoming the limitations of many current techniques:

1. It provides complete and precise descriptions of array designs, as well as sound reasoning for verifications.
2. It exploits all unique features of systolic arrays to produce notation and reasoning methods extremely suitable for systolic arrays, making specifications and verifications simple, elegant, and effective.
3. The constructs and operators used are aimed to be quite similar to those of Interval Temporal Logic (ITL, which is a powerful tool for lower level reasoning), and it can hence be unified with ITL to form a multilevel reasoning system without much difficulty. This extension is beyond the scope of the book.
4. It can deal with inherent parallelism and pipelining in an array and does not impose any restriction on array design flexibility.
5. Besides being useful for mathematical reasoning, it can also be applied to simulations, fault diagnoses, and test generations for array systems. Such applications are also beyond the scope of the book.
6. It can be automated and incorporated into CAD packages.

Moving in this direction, we developed a formalism for systolic array specifications and verifications. We call it *Systolic Temporal Arithmetic (STA)* [Ling89b, Ling90] since it is based on describing arithmetic operations in dynamic systolic environments. We further developed a verifier, named *VSTA*, to automate our verification process [Ling93, Ling95, Shih95]. VSTA is described in Chapter 6. We now discuss our abstraction mechanisms for producing this formalism.

## 2.5    ABSTRACTION MECHANISMS

Viewing hardware at different levels, a gap exists between the specification of the system at the higher level and that of the basic resources used in the implementation of the system (for example, the gap between the array level architecture and the PE, module, and circuit level architecture for systolic arrays). A large number of details are not represented at the higher levels. The specific mechanisms by which high-level descriptions abstract from lower-level details are called *abstraction mechanisms* [Evek85, MelT88, MelT93]. For systolic arrays, many of the PE and circuit level details that can be described by ITL are not represented at the array level. The abstraction mechanisms we adopt for our new formalism for array level specification and verification are summarized below:

1. **The Abstraction of Carriers:** In such mechanism hardware resources used in the implementation of the cells are abstracted. Hence the array level descriptions adopted do not include the description of implementation modules such as gates, registers, etc. These are described by ITL for lower level specification and verification purposes. Such abstraction frees implementers from any restrictions in implementing the cells.

2. **The Abstraction of Values and Operations:** ITL contains various useful operators and constructs for describing discrete time dynamic behavior of digital systems. This has been useful in many digital circuit specification and verification applications at the gate or circuit level. On the array level, however, the values of data and variables are more conveniently expressed in terms of real numbers instead of bit values 0 and 1. Moreover, arithmetic operations such as addition and multiplication are much more commonly involved instead of logical operations such as AND, OR, and others. We shall collectively call such abstractions as the abstraction of values and operations, which determines the syntax and the semantics of the new formalism. ITL specifications deal mainly with bit values and logical operations and are, in general, too tedious and complicated for reasoning at the array level.

3. **The Abstraction of Systolic Features:** Systolic arrays have some very nice and unique properties at the array level. They are synchrony, regularity, repeatability, modularity, spatial and temporal locality, pipelinability, and parallel processing ability. These properties are described in Section 1.2. The new formalism exploits these features in the following way:

- A new model of time is developed based on the fact that the entire array is controlled and synchronized by a global clock with fixed length clock cycles, and the fact that any changes of system state can only take place at the clock transitions and the system remains stable within each cycle.
- Due to the unique systolic array features, only a small set of operators and constructs are needed. For example, asynchronous features such as handshaking protocols and eventually operations (such as those similar to the eventually operator "$\Diamond$" of temporal logic) are not necessary due to the fact that the location and value of any data item or result at any given time can be precisely determined as a result of the highly synchronized property of systolic arrays. Having only a small set of constructs and operators greatly simplifies the usage of the formalism and allows CAD systems to have more flexibility in selecting efficient operators and constructs.
- New constructs are introduced into the notation to simplify lengthy expressions and to represent the unique systolic array properties. For example, universal quantifiers "$\forall$" and other high level constructs are introduced to describe the repeatable and regular connections of cells in a systolic array network. Constructs are also developed to describe features such as pipelinability, parallelism, and temporal locality.
- Systematic and efficient verification and reasoning procedures exploiting the repeatability and spatial locality features of systolic arrays are developed to speed up reasoning processes. For example, spatial locality enables us to propagate a reasoning process from cell to cell until the output boundary of the array is arrived. At each stage of the process, only the communication to/from the neighboring cells needs to be considered. This simplifies reasoning processes. Moreover, the repeatability, regularity, and locality nature of systolic arrays enable us to consider methods based on principles such as mathematical induction to speed up reasoning.

A complete description of our formalism, STA, is given in the next chapter, with techniques and framework for STA-based specification and verification given in Chapter 4, and application examples in Chapter 5. Chapter 6 presents our Prolog-based verifier designed to automate verification processes and Chapter 7 discusses how our techniques are used to verify a complex array designed for LU decomposition. A sample user interface and a sample verification session are provided in Appendix A.

# Chapter 3

# SYSTOLIC TEMPORAL ARITHMETIC: A FORMALISM

## 3.1    INTRODUCTION

In this chapter, the new formalism developed, named Systolic Temporal Arithmetic (STA), is formally introduced. The model of this formalism $L_{STA}$ is a 3-tuple:

$$L_{STA} = <C, D, M>$$

where $C$ is a non-empty set of cycles (or states) as defined by the systolic temporal frame $T_s$, discussed in Section 3.2 (STA time model), $D$ is a non-empty set of legal sentences composed by legal symbols and values, and $M$ is a mapping which assigns to each sentence in $D$ and each cycle in $C$, an element of $D$. From this model, STA axioms, rules, and theorems are developed to exploit systolic properties and to simplify and facilitate reasoning; these are illustrated and proved in Section 3.6. Higher level constructs that capture systolic array properties to simplify specification and verification are also developed and described in this chapter.

## 3.2    THE MODEL OF TIME

The model of time for STA is linear, discrete, and "systolic". We first define the concept of *temporal frame* T and then introduce *systolic temporal frame*, $T_s$, which serves as the model of time for STA and STA *cycles* $\in C$:

**Definition:**  A *temporal frame* T for a set of formalisms $L_T$ (for which $L_{STA}$ is a subset) is a 3-tuple:

$$T = <T_T, R, M>$$

where
$T_T$ is a non-empty set of time points,
R is a set of temporal precedence relations, and
M is a meaning function, which assigns to each sentence in $L_T$, its values throughout time.

The set of formalisms, $L_T$, can be defined by a 3-tuple model

$$<T_t, D, M>$$

where
$T_t$ is a non-empty set of time points or group of time points satisfying the constraints imposed by T,
D is a non-empty set of legal sentences composed of legal symbols and values defined by the syntax of $L_T$, and
M is a mapping, which assigns to each element in D and each element in $T_t$, an element of D.

**Definition:** A *systolic temporal frame* $T_s$ for $L_{STA}$ is a subset of the temporal frame T, which satisfies systolic constraints. More precisely, $T_s$ can be viewed as a 3-tuple:

$$T_s = < T_s, R, M >$$

where $T_s$, $R$, and $M$ are subsets of $T_T$, R, and M respectively, satisfying the following systolic constraints:

1. *Constraints on temporal precedence $R$:* For time points $t, s, r \in T_s$, the following temporal precedence constraints need to be satisfied:

$$\text{Transitivity: } \forall t \forall s \forall r. \quad R(t,s) \wedge R(s,r) \rightarrow R(t,r)$$
$$\text{Linearity: } \forall s \forall t. \quad R(s,t) \vee (s=t) \vee R(t,s)$$

   where $R(t, t')$ expresses the temporal precedence of $t$ over $t'$; "$\wedge$" indicates an AND relation and "$\vee$" refers to an OR relation. The above constraints simply characterize the linear time properties [Turn84] of $T_s$. Transitivity needs little justification; it simply states that whenever $t$ happens before $s$ and $s$ happens before $r$, then $t$ happens before $r$. Figure 3.1a depicts a transitive relation. Linearity constraint rules out branches in temporal past and future of the form shown in Figure 3.1b. Linear time insists on there being only one past and one future.

2. *Constraints on time points:* In interval temporal logic (ITL) [Mosz83, Mosz85], time points are finite sequences of states, which are referred to as intervals. An interval of time contains a non-empty, finite sequence of states. In systolic time frame, we use the interval/state concept as defined by ITL [Mosz83, Mosz85] with the following extensions (as shown in Figure 3.2):

   (a) We assign a clock cycle to a state; time points are therefore finite sequences of *cycles*

$$c_1, c_2, ..., c_n$$

   as shown in Figure 3.2. All cycles are of the same fixed duration.

   (b) An *interval* is defined as the time between loading the first input and unloading the last output into/from one array. Thus if problem instances require a computation time of $N$ cycles, the *interval* is said to consist of $N$ cycles.

(c) Under the constraints of systolic abstractions, changes of system state can only take place at the clock transitions (the boundary between $c_i$ and $c_{i+1}$, for all $i$'s), and system remains stable within each cycle; no changes are allowed within a *cycle*. The cycle is therefore the smallest *indivisible* time unit for systolic array description. In such a time frame, the "next" concept for temporal operators refers to the next "cycle".

(d) The first occurrence of the cycle $c_1$ (i.e. systolic array begins its operation) begins at time = 0.

3. *Constraints on function M for STA*: The meaning function $M$ is a function that maps each legal sentence in STA and each cycle $c_i \in C$ to a legal value in STA. Legal sentences and values of STA are those defined by STA syntax (Section 3.3). The meanings of the sentences in STA due to the imposing of systolic time are defined by using $M$.

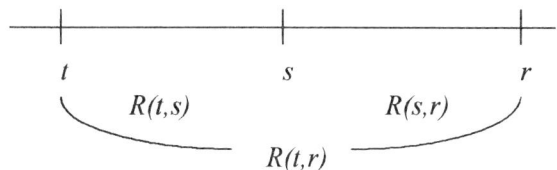

(a) A Transitive Relation

Branches in the past or in the future are not allowed in linear time

Linear time examples

(b) Temporal Linearity

Figure 3.1 Temporal transitivity and linearity

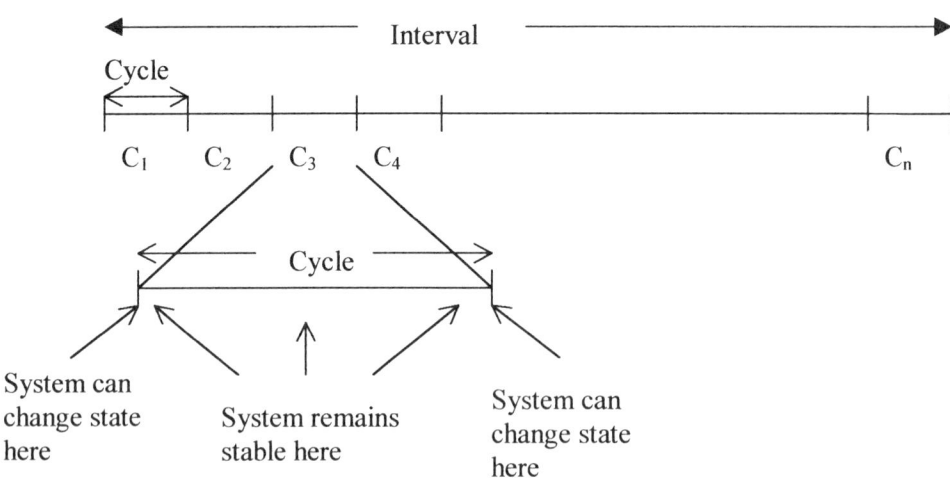

Figure 3.2 Model of time for STA

## 3.3    SYNTAX OF STA

The sentences in STA ($\in D$) are formed according to the syntax described in this section. The Backus-Naur Form (BNF) [Marc86] is adopted to specify the context-free syntax of STA. The meta-symbols used are:

| | | |
|---|---|---|
| ::= | | meaning "is defined as", same as $\equiv_{def}$ |
| \| | | meaning "or", used to separate alternatives |
| < > | | angle brackets used to surround category names |
| [ ] | | square brackets used to enclose optional items |
| { } | | braces used to indicate repetition of enclosed items, an arbitrary number (including null) of times |
| ( ) | | large parenthesis used to group elements of a syntactical unit. (Not to be confused with a regular-sized parenthesis "( )", which can be part of an STA sentence) |

Using the above meta-symbols, the context-free syntax of STA is presented as follows:

    <specification> ::= <structural spec.> <behavior spec.>

<structural spec.> ::= { $\models$ [<quantification>] <predicate> }

<behavior spec.> ::= <functional spec.> <input/output spec.>

<functional spec.> ::= { $\models$ [<quantification>] <implication> }

<input/output spec.> ::= { [<quantification>] <value predicate> }

<spec. sentence> ::= $\Big(\models$ [<quantification>]<predicate>$\Big)$ |

$\Big(\models$ [<quantification>] <implication>$\Big)$ |

$\Big($[<quantification>] <value predicate>$\Big)$

<implication> ::= <pred. or value pred.> { $\wedge$ <pred. or value pred.> }

$\rightarrow$ <value predicate> { $\wedge$ <value predicate>}

<quantification> ::= { [<quantifier>] [<integer>|<lower case letter>]
[<relational operator>] <lower case letter> [<relational operator>]
[<integer>|<lower case letter>] }

<pred. or value pred.> ::= <predicate> | <value predicate>

<predicate> ::= <name> (<subj. or iden.> [, <subj. or iden.>])

<subj. or iden.> ::= <subject> | <identifier>

<value predicate> ::= Val ( <identifier> , <general data> )

<name> ::= <upper case letter> {<lower case letter>}

<quantifier> ::= $\exists$ | $\forall$

<identifier> ::= <identifier name> ( <subject> , <subject> )

<identifier name> ::= Row | Col | In | Out | Mem

<subject> ::= <upper letter> {[<unsigned integer>]} [<index>]

<temporal equation> ::= <general data> = <general data>

<general data> ::= <temporal variable> | <temp. var. expr.> |

<temporal constant> | <temporal arith. expr.>

<temporal arith. expr.> ::= <temporal term> |

$\big($<sign> <temporal term>$\big)$ |

$\big($<temporal arith. expr.> <arith. op.> <temporal term>$\big)$

<temporal term> ::= <temporal data> |

$\big($<temporal term> <arith. op.> <temporal data>$\big)$

<temporal data> ::= <temporal variable> | (<temp. var. expr.>) |

(<temporal constant>) | (<temporal arith. expr.>)

<temporal variable> ::= <u>lower case letter</u> [<index>]

<temp. var. expr.> ::= $\big($<normal data> {,<normal data> }

,□ <normal data> $\big)$ | (□<normal data>$\big)$

<temporal constant> ::= $\big($<normal constant> {,<normal constant> }

,□ <normal constant> $\big)$ | (□<normal constant> $\big)$

<normal data> ::= <normal variable> | <normal constant> | (<normal arith. expr.>)

<normal arith. expr.> ::= <normal term> | $\big($<sign> <normal term>$\big)$ |

$\big($<normal arith. expr.> <arith. op.> <normal term>$\big)$

<normal term> ::= <normal factor> | $\big($<normal term> <arith. op.> <normal factor> $\big)$

<normal factor> ::= <normal variable>|<unsigned normal constant>|(<normal arith. expr.>)

<normal variable> ::= <lower case letter>[<index>]

<normal constant> ::= <number> | <unknown>

<unsigned normal constant> ::= <unsigned number> | <unknown>

\<index\> ::= \<ind.\> [, \<ind.\>]

\<ind.\> ::= $\Big([\{$\<lower case letter\> $|$ \<greek letter\>$\}]$ [\<number\>]$\Big)$ $|$

$\Big([$\<number\>$]$ $[\{$\<lower case letter\> $|$ \<greek letter\>$\}]\Big)$ $|$ \<index arith. expr.\>

\<index arith. expr.\> ::= \<ind. term\> $|$ $\Big($\<sign\> \<ind. term\>$\Big)$ $|$

$\Big($\<index arith. expr.\> \<arith. op.\> \<ind. term\>$\Big)$

\<ind. term\> ::= \<ind. data\> $\Big|$ $\Big($\<ind. term\> \<arith. op.\> \<ind. data\>$\Big)$

\<ind. data\>::=\<lower case letter\>|\<greek letter\>|\<unsigned number\>|

(\<index arith. expr.\>)

\<temporal operator\> ::= $\Box$ $|$ , $|$ $\bigcirc$ $^{[<ind.>]}$

\<arith. op.\> ::= $+$ $|$ $-$ $|$ $*$ $|$ $/$ $|$ div $|$ mod

\<logical operator\> ::= $\neg$ $|$ $\wedge$ $|$ $\vee$ $|$ $\oplus$ $|$ $\rightarrow$

\<relational operator\> ::= $=$ $|$ $\neq$ $|$ $>$ $|$ $\geq$ $|$ $<$ $|$ $\leq$

\<number\> ::= $\Big($\<integer\> [.\<unsigned integer\>]$\Big)$ $|$ \<number\> $^{[<number>]}$

\<unsigned number\> ::= $\Big($\<unsigned integer\> [.\<unsigned integer\>]$\Big)$ $|$

\<unsigned number\>$\}^{[<number>]}$

\<integer\> ::= [\<sign\>] \<unsigned integer\>

\<unsigned integer\> ::= \<digit\> {\<digit\>}

\<boolean\> ::= *true* $|$ *false*

\<sign\> ::= \<blank\> $|$ $+$ $|$ $-$

\<greek letter\> ::= $\alpha$ $|$ $\beta$ $|$ $\gamma$ $|$ $\delta$ $|$ $\varepsilon$ $|$ $\zeta$ $|$ $\eta$ $|$ $\theta$ $|$ $\iota$ $|$ $\kappa$ $|$ $\lambda$ $|$ $\mu$ $|$ $\nu$ $|$ $\xi$ $|$ $o$ $|$ $\pi$ $|$ $\rho$ $|$ $\sigma$ $|$ $\tau$ $|$ $\upsilon$ $|$ $\phi$ $|$ $\chi$ $|$ $\psi$
$|$ $\omega$

<upper case letter> ::= A | B | C | D | E | F | G | H | I | J | K | L | M | N | O | P | Q | R | S | T | U
| V | W | X | Y | Z

<lower case letter> ::= $a$ | $b$ | $c$ | $d$ | $e$ | $f$ | $g$ | $h$ | $i$ | $j$ | $k$ | $l$ | $m$ | $n$ | $o$ | $p$ | $q$ | $r$ | $s$ | $t$ | $u$ | $v$ | $w$
| $x$ | $y$ | $z$

<digit> ::= 0 | 1 | 2 | 3 | 4 | 5 | 6 | 7 | 8 | 9

<unknown> ::= _

<blank> ::=

We shall now discuss some unique features of STA syntax. Firstly, the distinction between *normal* and *temporal* variables and constants. A *normal variable* is a letter written in lower case italic font, with or without a subscript, each of which can have a value ranging over any real numbers. *Normal constants* are actual real numbers; a quantity with an unknown value is designated by " _ ". Each normal variable/constant is used to denote a value in only *one* cycle. In contrast, a *temporal variable* or a *temporal constant* has a value, which is a sequential temporal connection of real numbers from the first to the last cycle. A temporal variable, denoted by an underlined letter, written in lower case font, with or without a subscript, can be expressed by a sequential connection of normal variables, normal constants, and/or normal arithmetic expressions (syntactic category <temp. var. expr.>). For example, a temporal variable

$$\underline{u} = u_0, u_1, u_2, \square\, u_k$$

means that $\underline{u}$ is $u_0$ in the present cycle, $u_1$ in the next cycle, $u_2$ in the next next cycle, and $u_k$ thereafter. We use coma "," to separate variables in different cycles.

The canonical form of a temporal variable is

$$\underline{u} = u_0, u_1, u_2, \ldots, \square\, u_k$$

Here "," serves as temporal connection, and "$\square$" is a henceforth operator. To further simplify the writing of consecutive variables, the symbol

$$\bigwedge_{\alpha=a}^{b}$$

is created and is defined as

$$\bigwedge_{\alpha=a}^{b} u_\alpha \equiv_{def} u_a, u_{a+1}, \ldots, u_b$$

Hence the above canonical form can be written as

$$\underline{u} = \bigwedge_{\alpha=0}^{k-1} u_\alpha, \square u_k$$

Above is the most general form of expressing temporal variables. Any other form will be a special case of this form. For example, a temporal variable

$$\underline{u} = \square\, u_0$$

is a special case of this general form with $k = 0$. Similarly, a *temporal constant* is a sequential connection of normal constants.

Conventional *logical operators* and *arithmetic operators* are also used in STA. The *AND* operator "∧" is often used to provide temporal connection of variables. *Temporal operators* include *henceforth* "□" and *next* ",". They are used to operate on arithmetic operands to indicate arithmetic operations and values at different times. The priorities of the operators are implied in the BNF syntax description. Conventional mathematical operators have their priorities as in conventional mathematics.

*Identifiers* in STA are used to identify subjects with respect to subjects. The list of predefined identifiers is given below:

Row($i,S$) : Row $i$ of systolic array $S$
Col($i,S$) : Column $i$ of systolic array $S$
In($I_i$, $B_j$) : the input terminal $I_i$ of component $B_j$
Out($O_i$, $B_j$) : the output terminal $O_i$ of component $B_j$
Mem($M_i$, $B_j$) : the memory location $M_i$ of component $B_j$

*Predicates* are also used in STA; a list of predefined predicates is given below:

Sys(S) : S is a systolic array
Cell($B_i$) : $B_i$ is a cell (PE)
Conn($X, Y$) : $X$ and $Y$ are directly connected (by wires) (same as Conn($Y, X$)) (here $X$ and $Y$
          can be identifiers)
Val($X, m$) : the value of the data on/in the terminal/memory location $X$ is $m$

The last predicate is the *value predicate*, which is used to indicate data items on input/output/memory of a cell/array. Besides these predicates, users can define additional predicates if needed.

A complete systolic array specification consists of a sequence of STA specification sentences, each of which is in one of the three legal forms:

⊨ [<quantification>] <predicate>

⊨ [<quantification>] <implication>

[<quantification>] <value predicate>

and each should be written beginning from a new separate line to prevent confusions since we do not have explicit sentence separators in STA.

The legal values of STA include real numbers, temporal sequences of real numbers, boolean, and an unknown real value "_".

For reasoning at the array level, many bit level operators used in ITL, such as bit logical operators, operators for stable, rising, and falling signals, temporal assignment and temporal blocking, are not necessary. Bit level properties are dealt with at the lower levels.

## 3.4    SEMANTICS OF STA

A sentence in STA can be mapped to either a truth value, or a real number, or a temporal sequence of real numbers (e.g. temporal variables and constants), depending on the domain and the situation to which it is applied. We can make this statement more precise by stating the existence of cycles $c_1$, $c_2$, ..., $c_n$ and a meaning function $M$ that maps sentences and cycles to data values in data domain $D$. For example, if $x$ is a numerical variable,

$$M_{c_i} < x > = 8$$

signifies that the value of $x$ in cycle $c_i$ is 8.

The value of a sentence u on an interval or subinterval $I = c_1, c_2, ..., c_n$ is defined as the value of u in the initial cycle of $I$:

$$M_{c1,c2,...,cn} < u > \equiv_{def} M_{c1} < u >$$

All sentences used in the reasoning of a systolic array are by default defined over the interval specified for the array.

The semantics of arithmetic operators, relational operators, logical operators, quantifiers, functions, and predicates are well established and defined in conventional mathematics, propositional logic, and predicate logic, for static environments. We shall now extend these semantics to dynamic environments, letting u and v be sentences, arithmetic operands, etc., in STA:

1. *Arithmetic, logical, or relational operations.* The semantics for operations on operands over an interval are:

$$M_{c1,c2,...,cn} < op \ u > \equiv_{def} op \ M_{c1,c2,...,cn} < u >$$
$$M_{c1,c2,...,cn} < u \ op \ v > \equiv_{def} M_{c1,c2,...,cn} < u > op \ M_{c1,c2,...,cn} < v >$$

where *op* is an unary or a binary operator. For instance,

$$M_{c1,c2,\ldots,cn} < \neg \, u > \equiv_{def} \neg \, M_{c1,c2,\ldots,cn} < u >$$

$$M_{c1,c2,\ldots,cn} < u + v > \equiv_{def} M_{c1,c2,\ldots,cn} < u > + \, M_{c1,c2,\ldots,cn} < v >$$

$$M_{c1,c2,\ldots,cn} < u = v > \equiv_{def} M_{c1,c2,\ldots,cn} < u > = M_{c1,c2,\ldots,cn} < v >$$

2. *Quantifiers.* The semantics of quantifiers over an interval are defined as follows:

$$M_{c1,c2,\ldots,cn} < \forall i. \, u > \equiv_{def} M_{c1',c2',\ldots,cn'} < u > \quad \text{for any } \alpha$$

$$M_{c1,c2,\ldots,cn} < \exists i. \, u > \equiv_{def} M_{c1',c2',\ldots,cn'} < u > \quad \text{for some } \alpha$$

where $c_1', c_2', \ldots, c_n' = (c_1, c_2, \ldots, c_n) \, [i/\alpha]$. Here "/" denotes a substitution, $i$ is a variable and function constant $\alpha$ maps variables to values in data domain.

3. *Predicates.* The meaning of a predicate $P(y_1, y_2, \ldots, y_k)$ is given by

$$M < P > \in (D^k \to \{true, false\})$$

where $D^k$ denotes the Cartesian product of $k$ copies of $D$ and "$\to$" denotes a mapping here, it can be extended to take into account of dynamic behavior over time as

$$M_{c1,c2,\ldots,cn} < P(y_1, y_2, \ldots, y_k) > \equiv_{def} M < P \, ( M_{c1,c2,\ldots,cn} < y_1 >, M_{c1,c2,\ldots,cn} < y_2 >, \ldots, M_{c1,c2,\ldots,cn} < y_k >) >$$

Functions can be similarly defined. Functions are high level constructs developed by users to capture collective arithmetic operations.

4. *Temporal operators.* Unlike temporal logic, temporal operators in STA mainly operate on arithmetic entities instead of on logical entities. Their semantics are given as follows:

$$M_{c1,c2,\ldots,cn} < u > \equiv_{def} M_{c1} < u >$$

$$M_{c1,c2,\ldots,cn} < \square u > \equiv_{def} \forall i \quad 1 \le i \le n. \quad M_{ci} < u >$$

$$M_{c1,c2,\ldots,cn} <, u > \equiv_{def} M_{c2} < u > \qquad (n \ge 2)$$

Normal variables and normal constants have their normal meanings and functions as in conventional mathematics, each can have a real number value. Identifiers and subjects are explained in Section 3.3. As for the other STA constructs, each of them is simply the combination of the operations discussed in this section; hence their semantics are simply the combination of one or more of the already defined semantics. For example, a temporal variable expression $\underline{u} = u_0, u_1, \square \, u_2$ is defined as

$$M_{c1,c2,\ldots,cn} < u_0, u_1, \square u_2 > \equiv_{def} M_{c1} < u_0 > \text{ and } M_{c2} < u_1 > \text{ and } \forall i. \, 3 \le i \le n. \quad M_{ci} < u_2 >$$

## 3.5  SATISFIABILITY AND VALIDITY

If a particular sentence u is true on an interval $c_1, c_2, ..., c_n$ (or a cycle $c_i$), we say that the interval (or cycle) *satisfies* the sentence, written as

$$c_1, c_2, ..., c_n \models u \text{ (for intervals)} \quad \text{or} \quad c_i \models u \text{ (for cycle)}$$

If a sentence is satisfied by all cycles and intervals, it is said to be *valid*, written as

$$\models u$$

Valid sentences describe properties in all situations and are essentially theorems about temporal behavior.

## 3.6  STA AXIOMS, RULES AND THEOREMS

### 3.6.1  Axioms

Let $u_0, u_1, ..., u_k, v_0, v_1, ..., v_m, x,$ and $y$ be normal variables, some basic axioms in STA for which the STA rules are based upon are introduced as follows:

**Axiom 1:** Let "*op*" be a binary arithmetic, logical, or relational operator operates on two temporal variables, then quantities of the same cycle are "*op*" with each other; that is

$$(\bigwedge_{i=0}^{k-1} u_i, \square\, u_k)\, op\, (\bigwedge_{i=0}^{m-1} v_i, \square\, v_m) = \bigwedge_{i=0}^{m}(u_i\, op\, v_i), \bigwedge_{i=m+1}^{k-1}(u_i\, op\, v_m), \square\,(u_k\, op\, v_m) \qquad \text{(if } k \geq m)$$

or

$$\bigwedge_{i=0}^{k}(u_i\, op\, v_i), \bigwedge_{i=k+1}^{m-1}(u_k\, op\, v_i), \square\,(u_k\, op\, v_m) \qquad \text{(if } k \leq m)$$

**Axiom 2:** Let "*op*" be a binary arithmetic operator, then

$$x\, op\, \_ = \_\, op\, x = \_$$

That is, if one operand of "*op*" is unknown, then the result is unknown. If $x = 0$ for multiplication, then

$$0 * \_ = \_ * 0 = 0$$

**Axiom 3:** Quantities that appeared in time $< 0$ are purely mathematical results; they have no practical meaning and can be ignored.

**Axiom 4:** Two temporal quantities are equal if and only if their quantities are equal in each cycle:

$$(\bigwedge_{i=0}^{k-1} u_i, \Box u_k) = (\bigwedge_{i=0}^{m-1} v_i, \Box v_m)$$

$$\text{iff} \quad (\forall i \quad 0 \le i \le m. \quad (u_i = v_i)) \wedge (\forall i \quad m+1 \le i \le k. \quad (u_i = v_m)) \qquad (\text{if } k \ge m)$$

$$\text{or}$$

$$(\forall i \quad 0 \le i \le k. \quad (u_i = v_i)) \wedge (\forall i \quad k+1 \le i \le m. \quad (u_k = v_i)) \qquad (\text{if } k \le m)$$

The validity of these axioms can be seen trivially using common sense reasoning and arithmetic. These axioms are used frequently in our reasoning using STA.

## 3.6.2 Rules

Based on these axioms, several STA rules can be derived to be effectively utilized for systolic array reasoning. Additional rules for other arithmetic and logical operations can be developed if needed.

**Rule 1.** Henceforth "$\Box$" rule:

$$\Box x = x, \Box x$$

**Rule 2.** Delay addition rule:

$$\bigwedge_{r=0}^{p} \_, (\bigwedge_{r=0}^{q} \_, x) = \bigwedge_{r=0}^{p+q+1} \_, x$$

**Rule 3.** Delay rule:

$$\_, (\bigwedge_{i=0}^{k-1} u_i, \Box u_k) = (\_, \bigwedge_{i=0}^{k-1} u_i, \Box u_k)$$

These rules follow directly from temporal logic [Abad86, Hail80]. The rules described in this section, plus other basic rules in mathematical operations and conventional logical reasoning, can be used for simplifications, deductions, and often provide convenient and natural shortcuts in proofs and reasoning for many systolic arrays. Users of STA can develop more rules if needs occur.

### 3.6.3  Theorems

Two theorems, which are particularly useful in the reasoning of bi-directional systolic arrays, are developed as follows:

**Theorem 1:** Let $\underline{u}_1$, $\underline{u}_i$, $\underline{u}_{i+1}$, etc., denote temporal variables, then the STA difference equation

$$\underline{u}_i = \_, \underline{u}_{i+1} \qquad\qquad \text{for } i = 0, 1,..., k\text{-}1 \qquad\qquad (\text{T1.1})$$

has solution

$$\underline{u}_t = \overset{k-t-1}{\underset{r=0}{\bigwedge}} \_, \underline{u}_k \qquad\qquad \text{for } t = 0, 1,..., k \qquad\qquad (\text{T1.2})$$

**Proof:**   This can be proved by successive substitution by Equation (T1.1) and the successive use of STA Rules (2) and (3), starting from Equation (T1.1) with $i = t$:

$$\underline{u}_t \;\; = \;\; \_, \underline{u}_{t+1}$$

$$= \_, \_, \underline{u}_{t+2}$$

$$= \_, \_, \_, \underline{u}_{t+3}$$

$$= \_, \_, \_, ..., \underline{u}_k$$

$$= \overset{k-t-1}{\underset{r=0}{\bigwedge}} \_, \underline{u}_k$$

which is Equation (T1.2) for $t = 0, 1,..., k\text{-}1$. For $t = k$, the correctness of Equation (T1.2) is trivial. Hence $\underline{u}_t$, for $t = 0, 1, ... , k$, is given by Equation (T1.2).                                      □

**Theorem 2:** Let $\underline{u}_i$, $\underline{v}_i$, etc., denote temporal variables, then the STA difference equation

$$\underline{v}_{i+1} \;\; = \;\; ( \_, \underline{v}_i ) + \underline{u}_i \qquad\qquad \text{for } i = 1, 2, ..., k \qquad\qquad (\text{T2.1})$$

has solution

$$\underline{v}_t \;\; = \;\; ( \overset{t-2}{\underset{r=0}{\bigwedge}} \_, \underline{v}_1 ) + \sum_{j=1}^{t-1} ( \overset{j-2}{\underset{r=0}{\bigwedge}} \_, \underline{u}_{t-j} ) \qquad\qquad \text{for } t = 2, 3, ..., k+1 \qquad\qquad (\text{T2.2})$$

**Proof:**   This can be proved by induction:

*Basis of Induction:* For $i = 1$ in Equation (T2.1), we have

$$\underline{v}_2 \;\; = \;\; ( \_, \underline{v}_1 ) + \underline{u}_1$$

which is Equation (T2.2) for $t = 2$.

*Inductive Hypothesis*: Assume that for any $t = 2, 3,..., k$, $\underline{v}_t$ is given by Equation (T2.2).

*Inductive Step:* We now derive the expression for $\underline{v}_{t+1}$. From Equation (T2.1) we have

$$\underline{v}_{t+1} \quad = \; ( \_, \underline{v}_t) + \underline{u}_t$$

$$= \; ( \_, ((\bigwedge_{r=0}^{t-2} \_, \underline{v}_1 ) + \sum_{j=1}^{t-1} (\bigwedge_{r=0}^{j-2} \_, \underline{u}_{t-j}))) + \underline{u}_t$$

By STA Axioms (2) and (3) as well as Rules (2) and (3), this is

$$= ( \_, \bigwedge_{r=0}^{t-2} \_, \underline{v}_1 + \sum_{j=1}^{t-1} (\bigwedge_{r=0}^{j-1} \_, \underline{u}_{t-j})) + \underline{u}_t$$

$$= \_, \bigwedge_{r=0}^{t-2} \_, \underline{v}_1 + \sum_{j=0}^{t-1} (\bigwedge_{r=0}^{j-1} \_, \underline{u}_{t-j})$$

$$= \bigwedge_{r=0}^{t-1} \_, \underline{v}_1 + \sum_{j=1}^{t} (\bigwedge_{r=0}^{j-2} \_, \underline{u}_{t-j+1})$$

which proves that $\underline{v}_{t+1}$ is given by Equation (T2.2).

*Conclusion:* Hence $\underline{v}_t$, for $t = 2, 3, ..., k+1$, is given by Equation (T2.2). □

## 3.7   USEFUL CONSTRUCTS

Besides synchronization and value/operation abstraction, many features of systolic arrays can also be captured by introducing new constructs into STA notation. These constructs simplify specifications, facilitate reasoning, and help users in describing large arrays or arrays of a parameterized size. *Functions* can also be defined by users to capture collective arithmetic operations. Note also that for bounded quantification, the form "$\forall 1 \leq i \leq n$" carries the same meaning as "$\forall i. \; 1 \leq i \leq n$" and both forms are used interchangeably in the book.

**1.  Expressing Repeatability, Regularity, and Spatial locality:**

(a) *Constructs for Identical Components:* Universal quantifier "$\forall$" and vector notation (variables in bold font) are introduced to specify repeated identical cells, as well as regular and local interconnections in systolic arrays to simplify specifications. For example, a structural specification for an array with $n$ identical cells (PEs), $B_1$ to $B_n$, can be specified as:

$$\models \ \forall \ 1 \leq i \leq n. \quad Cell(B_i)$$

We define a vector **B** with elements $B_1$, $B_2$, ..., $B_n$, and define

$$Cell(\mathbf{B}) \equiv_{def} \ \forall \ 1 \leq i \leq n. \quad Cell(B_i)$$

We then specify these cells collectively as $\models Cell(\mathbf{B})$.

(b) *Constructs for Connections:* The construct "Conn" can be used to specify regular and local connections of cells in systolic arrays. An example is

$$\forall i. \quad 1 \leq i \leq n-1. \quad Conn \ (Out(O_\alpha, B_i), In(I_\alpha, B_{i+1}))$$

where the local connections of cells are specified. The output $O_\alpha$ of each cell is connected to the input $I_\alpha$ of its neighboring right cell, as shown in Figure 3.3.

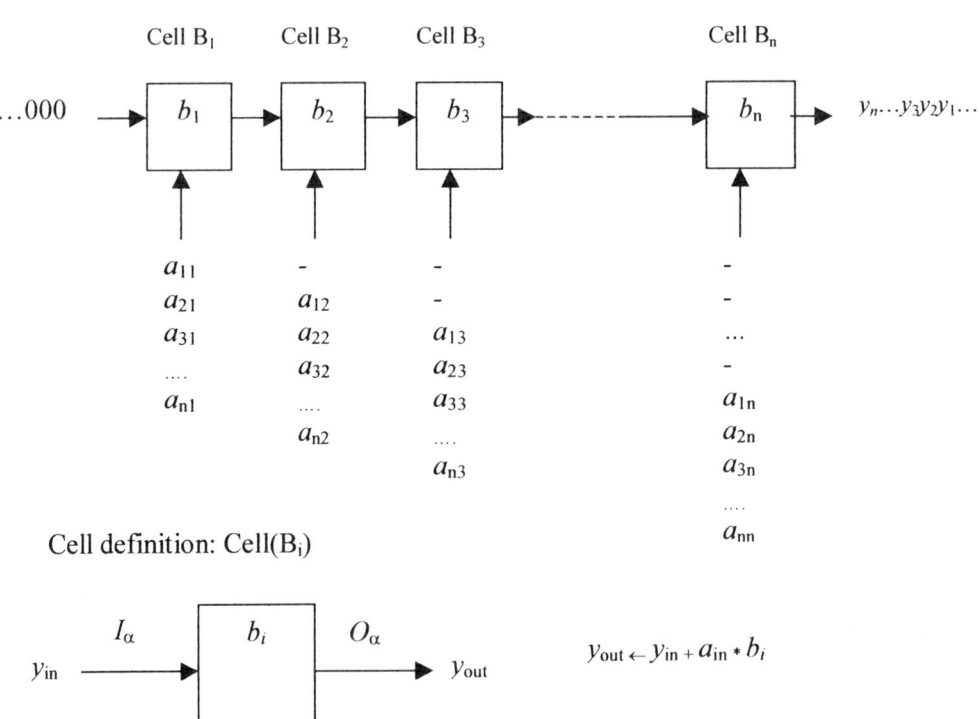

Figure 3.3 1-D systolic array for matrix-vector multiplication and its cell definition

(c) *Constructs for Components' Functions*: Cell functions are specified by implications. An example of specifying the cell function for the array of Figure 3.3 is given below.

$$\models \quad \forall B_i \quad \mathrm{Cell}(B_i) \wedge \mathrm{Val}(\mathrm{In}(I_\alpha, B_i), y_{in}) \wedge \mathrm{Val}(\mathrm{In}(I_\beta, B_i), a_{in}) \wedge \mathrm{Val}(\mathrm{Mem}(M, B_i), b_i)$$
$$\rightarrow \mathrm{Val}(\mathrm{Out}(O_\alpha, B_i), \_, (y_{in} + a_{in} * b_i))$$

A special case of a component is a direct connection (wires), whose function is simply to transmit signals from one end to the other without alterations. This is specified as

$$\models \quad \forall X \forall Y \quad \mathrm{Conn}(X, Y) \wedge \mathrm{Val}(X, x) \wedge \mathrm{Val}(Y, y) \rightarrow \mathrm{Val}(Y, x)$$

## 2. Expressing Pipelinability, Parallelism, and Temporal locality

The nature of temporal locality of a systolic array is inherent in its pipelining nature and is perfectly captured by operators such as " ," and " $\wedge$ " in STA sentences. The construct $\bigwedge\limits_{\alpha=a}^{b}$, defined as

$$\bigwedge_{\alpha=a}^{b} u_\alpha \equiv_{def} u_a, u_{a+1}, \ldots, u_b$$

is used to capture successive values as a result of pipelining in systolic arrays to simplify temporal variable expressions.

As an example, the output $O_\alpha$ of the systolic array of Figure 3.3, as a result of pipelining, can be expressed as

$$\mathrm{Val}(\mathrm{Out}(O_\alpha, B_n), \bigwedge_{r=0}^{n-1} \_, \bigwedge_{r=1}^{n} (\sum_{i=1}^{n} a_{ri} * b_i), \Box\_)$$

To capture successive similar variables or quantities, each separated from the other by one cycle time, the construct $\bigwedge\limits_{\alpha=a \text{ by } 2}^{b}$, defined as

$$\bigwedge_{\alpha=a \text{ by } 2}^{b} (u_\alpha, v_\alpha) \equiv_{def} u_a, v_a, u_{a+2}, v_{a+2}, u_{a+4}, v_{a+4}, \ldots, u_b, v_b$$

is introduced. An example is given in Chapter 5.

For parallel computation in rows or columns, Bounded quantifiers such as $\forall i, 1 \leq i \leq n$ can be used to express them. Note that bounded quantifier is also used to describe structural and functional specifications, as discussed earlier.

This concludes our description of STA. Its applications to systolic array design begin from the next chapter.

# Chapter 4

# SPECIFICATION AND VERIFICATION FRAMEWORK

## 4.1  SYSTOLIC ARRAY SPECIFICATION

A formal specification of a systolic array using STA is a mathematical description of its design structure and its intended behavior at the array architectural level. STA notation and a good specification framework constitute a precise and complete description of a systolic array. A detailed formal STA syntax of the specification framework described in BNF is given in Section 3.3. An informal description is provided in this section. Specific examples are provided in Chapters 5 and 7.

A complete specification of a systolic array include:

1. **Structure specification,** which consists of the specifications of:

   (a) **Types of components in array:** Here the array and the types of cells (PEs) and modules involved are specified. Presented in BNF, the specification takes the form

   $$\{[<quantification>]\ <predicate>\}$$

   The type of cells or modules used are specified in predicates, while the number of each is described by the quantification construct using quantifiers. Details of these forms are given in Section 3.3.

   (b) **Structural connectivity**: Here the structural manner in which cells are connected to form the entire systolic network is specified. Presented in BNF, the specification takes the form

   $$\{[<quantification>]\ <predicate>\}$$

   Here each <predicate> indicates the interconnection predicate "Conn" (Section 3.7) used and identical connection patterns are captured by the quantification construct using quantifiers. Examples of structural specification are given in the next chapter.

2. **Behavior specification,** which consists of the specifications of:

(a) **Function of each component and connection:** Here the function of each cell (PE) or module used in composing the array is specified. The function of the wire connections is also specified here (the function of wire connection is trivial, it simply transmits signals from one end to the other without changes). These functions are specified using implications, described in BNF as

$$\{[<quantification>] <implication>\}$$

The antecedent of each implication is the conjunctions of several predicates representing the type of component (the type of PE, module, etc.) and the inputs and/or memory contents of this component. The consequent of each implication is a value predicate (Val) or the conjunctions of several value predicates representing the outputs and/or memory contents of this component as a result of the type of operation (depends on the type of component) been performed on its inputs and/or memory contents. Functions of identical components are captured by the quantification constructs using quantifiers. Details of these forms can be seen in Section 3.3 and the examples of Chapter 5.

(b) **Input behavior:** Here the intended dynamic pattern of the input data at each input of the systolic array must be specified. The preloaded data in each memory location should also be specified here. These are specified using value predicates, described in BNF as

$$\{[<quantification>] <value\ predicate>\}$$

Temporal variable expressions are usually used to express the input or preloaded memory data in these value predicates. Similar input patterns are captured by the quantification constructs using quantifiers.

(c) **Output behavior:** Here the *intended* dynamic pattern of the array outputs (or memory contents if these are treated as outputs) must be specified. Temporal variables expressing the output results should take the form of the mathematical specification at the algorithm level with some expected array operation cycle delays. The unknown constant "_" can be used whenever the value at certain time is unknown and of no importance. Output behavior specification takes the form (in BNF)

$$\{[<quantification>] <value\ predicate>\}$$

Temporal variable expressions are used to express these outputs in value predicates and similar output patterns can be captured by the quantification constructs using quantifiers.

Specifications 1(a), 1(b), and 2(a) above provide a complete physical description of the design at the array level while specifications 2(b) and 2(c) give a functional description of the array. Examples of these specifications are presented in the next chapter and the complete syntax is presented in Section 3.3. Higher level constructs can be used wherever convenient to simply specifications.

## 4.2  SYSTOLIC ARRAY VERIFICATION

A formal verification of a systolic array at the array architectural level is a mathematical process for checking whether the array level architecture realizes the algorithm level specification. The verification of a systolic array is the proving of the correctness of the following implication:

*<Structural specification>* ∧ *<Specification of functions of components and connections>* ∧ *<Input behavior specification>* → *<Output behavior specification>*     (4.1)

That is, the correctness of the array means that if the array is built according to the structural specification, if the array inputs satisfy the input specification (includes initial conditions), and each array component and connection operates in faithfulness to its functional specification, then the array outputs are guaranteed to satisfy the output specification. Since the output specification constitutes the algorithm level mathematical specification with appropriate array operation cycle delays, proving of the correctness of this implication will mean that the array realizes the algorithm level specification.

Verification can be performed in various ways, using forward reasoning [Kell82], backward reasoning [Ueha83], direct, refutation, or other methods. Three verification methods [Ling90, Ling95] are introduced in this chapter. These methods exploit the unique properties of systolic arrays to provide efficient proofs. They can be applied to a wide range of systolic arrays. Examples of such applications are discussed in the next chapter.

## 4.3  VERIFICATION BY OUTPUT DERIVATION AND COMPARISON

Verification by output derivation and comparison is the most direct way of verifying systolic array design. This method exploits the regularity and locality nature of systolic arrays. The method is described by the following steps:

1. **Output Derivation.** Using the given structural and behavioral specifications (except output behavior specification) as premises, together with the axioms, rules, theorems, and constructs of STA, derive the output behavior of the array by output propagation:

   (a) Derive the outputs of the cells nearest to the array inputs.

   (b) These outputs then become the inputs to the next level of cells and their outputs can be derived.

   (c) *If* the output behavior of the array has arrived

   *then go to* step 2;

   *else go to* step 1(b).

2. **Output Comparison.** Compare the output behavior derived in step 1(c) with the output behavior specification of the array:

   *If* the output of step 1(c) is *equal* to the output specification of the array

   *then* the correctness of the array is verified (due to correctness of implication (4.1));

   *else* the array is incorrect with respect to the algorithm level specification.

For an incorrect array, backward tracing or other techniques can be used to identify design faults. The output derivation and comparison method cannot be applied to certain arrays; for example, it cannot be applied to verify bi-directional systolic arrays, where data flow in two directly opposite directions; in such cases the method discussed in Section 4.5 is useful.


## 4.4    VERIFICATION BY INDUCTION

In cases where the size of the array to be verified is large, the previous method could take enormous time. We introduce the principle of mathematical induction for systolic array verification as an efficient reasoning method for STA. Although the concept of mathematical induction cannot be applied to verify many hardware systems, it is *interesting* to see that the concept is very suitable for many systolic array verifications. This is due to the repeatability, regularity, and locality nature of systolic arrays. The principle exploits these properties to construct a proof that an array of any size is correct. The method is especially useful for large arrays or arrays of parameterized sizes, since the number of steps in the procedure does not depend on the number of cells in general. The Principle of Mathematical Induction for Systolic Array Verification is stated as follows:

 Let $P(n)$ be the statement of the form

*<Structural specification>* ∧ *<Specification of functions of components and connections>* ∧ *<Input behavior specification>* → *<Output behavior specification>*

for a systolic array of size $n$ (can be $n$ cells, $n$ rows, etc.). There are three steps to proof:

1. **Basis of Induction.** Show that $P(n_0)$ is correct. ($n_0 < n$).

2. **Inductive Hypothesis.** Assume $P(k)$ is correct for $k \geq n_0$.

3. **Inductive Step.** Show that $P(k+1)$ is correct on the basis of the inductive hypothesis. That is, $P(k)$ implies $P(k+1)$.

Then we can conclude that $P(n)$ is correct for all $n \geq n_0$; and hence the array of size $n$ is correct with respect to its upper level specification.

As an example, consider the case of a linear systolic array with $n$ cells (size $n$) shown in Figure 3.3. Given the structural, component functional, and input behavioral specifications, we first show that the output of the cell $B_1$ is equal to the desired output specification for the array of size 1; we then assume that the output of some intermediate cell $B_k$ is equal to the output specification for the array of size $k$; if we can show that the output of the next cell $B_{k+1}$ is equal to the output specification for the array with size $k+1$, then we can validly conclude that the output of the last cell $B_n$ is equal to the output specification for the array of size $n$. Since the output of cell $B_n$ is actually the output of the given array of size $n$, the array is proved correct.

Although applying induction to proving 1-D arrays is not difficult, applying it to 2-D arrays and automating the proof with logic programming techniques are not as trivial. We expand our mathematical induction method to four induction techniques, introduced as follows.

## 1. Regular Mathematical Induction:

The regular mathematical induction procedure is stated as follows:

Let $P(n)$ be the statement of the form

> <Structural Specification> ∧<Component function Specification> ∧ <Input Specification>
> → <Output specification>

for a systolic array of size $n$ (can be $n$ cells, $n$ rows, etc.). There are three steps to proofs:

(a) *Basis of Induction*: Show that $P(n_0)$ is correct. $(n_0 < n)$.

(b) *Inductive Hypothesis*: Assume $P(k)$ is correct for any $k$, $k \geq n_0$.

(c) *Inductive Step*: Show that $P(k+1)$ is correct on the basis of the inductive hypothesis. That is, $P(k)$ implies $P(k+1)$.

Then we can conclude that $P(n)$ is correct for all $n \geq n_0$, and hence the array of size $n$ is correct with respect to its upper level specification. This can be used to prove a 1-D array such as that of Figure 5.1, for example.

## 2. Structured Induction:

The proof procedure is the same as mathematical induction except that in the *inductive hypothesis*, we assume $P(k)$ is correct for *all* $k$, $n > k \geq n_0$. This is a stronger induction.

## 3. Double Induction:

If $P(n)$ to be proved has more than one parameter to deal with (common for multi-dimensional array with several indices $i$, $j$, etc.), this is a technique of applying induction to more than just one parameter. If we know the right induction order, we can first choose one parameter, fix the other parameters and induce on this one. We can then apply the same approach to the other parameters one at a time, in a right order, until $P(k+1)$ is proved. This technique is necessary for proving many 2-D arrays, including the matrix-matrix multiplication example presented in Chapter 5 and in the appendix. In these cases, after selecting a right order to prove a 2-D array, we can perform induction on each dimension by applying mathematical, structured, or reverse induction to one parameter at a time.

### 4. Reverse Induction:

This is similar to the mathematical induction except that the proof is proceeded backward. Our *base case* consists of proving $P(N)$, where $N$ is a very large number. The *induction step* consists of proving $P(k-1)$ assuming $P(k)$. Then we have a proof for $P(n)$ for $n \leq N$. Such proof is sufficient in some cases. For example, suppose we apply double induction on two parameters. We can apply regular induction on one parameter, and reverse induction on the second parameter if the second parameter can be bounded in terms of the first one. In Chapter 7, we apply reverse induction in providing our sub-proof for the sub-array $S_k$ in the LU decomposition example of Figure 7.1. This technique is useful there because the sub-array is a trapezoid, forming a part of the triangular array as shown in Figure 7.1, with the longest row closest to the inputs. In this case we can prove the longest row first and use it as the base for induction.

Even though mathematical induction method greatly simplifies the steps involved in verifying large arrays or arrays of any parameterized sizes, it is difficult, or sometimes even impossible, to apply to cases where the array topologies and their data flow directions are multidimensional. In some of these cases, the method requires some human ingenuity; and in others, induction cannot be applied and the other two verification techniques may be more effective. Induction techniques cannot be applied to bi-directional systolic arrays, where data flow in two directly opposite directions; in such cases the method discussed in the next section is necessary.

## 4.5   VERIFICATION BY SOLVING STA DIFFERENCE EQUATIONS

Both the output derivation and comparison method (Section 4.3) and the induction method (Section 4.4) cannot be used to verify the correctness of bi-directional systolic arrays, having data flowing in two directly opposite directions. Cases similar to this with complex data flow may require a new method. Recalling that using difference equations to model cell computations were earlier proposed by Melhem and Rheinboldt [Melh84], we introduce the method of solving STA difference equations for systolic array verification as another efficient reasoning method for STA. This method exploits the regularity, locality, and repeatability nature of systolic arrays. Besides proving the correctness of bi-directional systolic arrays, the method can also be applied to some cases where the arrays are multidimensional or where the data flows are unidirectional or multidirectional. It can also be used for array of any parameterized size, since the number of steps in the procedure is independent of array size. The disadvantages of this method are, firstly, for simple arrays and data flows, it takes more time to verify an array as compared to using the other two methods. Secondly, it depends on our ability to solve the system of difference equations involved. Fortunately, in the cases of many systolic arrays, STA difference equations take the forms of Equations (T1.1) and (T2.1) of STA Theorems (1) and (2) (Section 3.6.3), which we already have developed the solutions. The procedure for systolic array verification by solving STA difference equations is stated as follows:

1. **STA Difference Equations Generation:** For a typical interior cell $i$ in the array, express each of its output behavior in terms of its inputs, using temporal variables. This can be obtained from the functional specification of the cell. From here we form a system of STA difference equations describing the I/O relation of each typical cell.

2. **Array I/O Relation Derivation:** This is done by solving the system of STA difference equations obtained from Step 1. The results are STA equations expressing each output of the array in terms of the array inputs.

3. **Output Behavior Derivation:** Substitute each input temporal variable in the output expressions of Step 2 by the corresponding normal variables and constants (obtained from the array input behavior specification) and manipulate the expressions using STA axioms and rules to obtain the output behavior of the array.

4. **Output Comparison:** Compare the output derived from Step 3 with the output specification of the array:

   *If* the output of Step 3 is *equal* to the output specification of the array

   *then* the correctness of the array is verified;

   *else* the array is incorrect with respect to the algorithm level specification.

Examples of applying all the three techniques to systolic array verifications are given in the next chapter.

**Chapter 5**

---

# SPECIFICATION AND VERIFICATION OF SYSTOLIC ARRAYS: APPLICATION EXAMPLES

## 5.1    A 1-D ARRAY FOR MATRIX-VECTOR MULTIPLICATION

Matrix-vector multiplication is a common step in many digital signal processing algorithms, an example of specifying and verifying a systolic array for implementing this is given in this section. The matrix-vector multiplication problem A * b for matrix A and vector b is defined as follows:

Given $b_1, b_2, ..., b_n$ and $a_{11}, a_{12}, ..., a_{1n}, a_{21}, a_{22}, ..., a_{nn}$

Compute $y_1, y_2, ..., y_n$

defined by $y_r = \sum_{i=1}^{n} a_{r,i} * b_i$

A linear 1-D systolic network for implementing matrix-vector multiplication can be produced as given in Figure 5.1 [KungS88]. All $y_r$'s are initialized to zeros and all $b_i$'s are preloaded into the cells. $a$'s inputs are skewed as shown in the figure. Computations are pipelined and the results start to appear at the output of cell $B_n$ $n$ cycles after the first input, followed by a new output every cycle. Total computation time is expected to be $2n$ cycles. The specification and the verification for this array are presented in the following subsections.

### 5.1.1  Array Specification using STA

Using vector notation **B** to indicate the cells $B_1$, $B_2$, ..., $B_n$ collectively, and using normal variables to denote the data elements, the specification of this systolic array of size $n$ in STA is given as follows:

1. **Structure Specification:**

    (a) **Type of components:**

(A1)  $\models$ Cell(**B**)        (i.e. $\models$ $\forall 1 \leq i \leq n.$ Cell($B_i$) )

**(b) Structural connectivity:**

(A2)  $\models$ $\forall 1 \leq i \leq n-1.$ Conn(Out($O_\alpha, B_i$), In($I_\alpha, B_{i+1}$))

## 2.  Behavior Specification:

**(a) Functions of individual components and connections:**

(B1)  The function of the cells used:

$$\models \quad \forall B_i \ \text{Cell}(B_i) \wedge \text{Val}(\text{In}(I_\alpha, B_i), y_{in}) \wedge \text{Val}(\text{In}(I_\beta, B_i), a_{in}) \wedge \text{Val}(\text{Mem}(M, B_i), b_i)$$
$$\rightarrow \text{Val}(\text{Out}(O_\alpha, B_i), \ \_, (y_{in} + a_{in} * b_i))$$

Here "*" denotes a multiplication operation.

The function of a connection (wire connection) is always implied and served as a rule in reasoning process:

$$\models \quad \forall X \forall Y. \ \text{Conn}(X, Y) \wedge \text{Val}(X, x) \wedge \text{Val}(Y, y) \rightarrow \text{Val}(Y, x)$$

**(b) Input behavior specification:**

(C1)  Val(In($I_\alpha, B_1$), $\square$ 0)

(C2)  $\forall 1 \leq i \leq n.$ Val(In($I_\beta, B_i$), $\bigwedge\limits_{r=0}^{i-2} \_, \bigwedge\limits_{r=1}^{n} a_{ri}, \square$ 0)

This is equivalent to

$$\text{Val}(\text{In}(I_\beta, B_1), \ a_{11}, a_{21}, a_{31}, \ldots, a_{n1}, 0, 0, \ldots, 0) \ \wedge$$
$$\text{Val}(\text{In}(I_\beta, B_2), \ \_, a_{12}, a_{22}, a_{32}, \ldots, a_{n2}, 0, 0, \ldots 0) \ \wedge$$
$$\text{Val}(\text{In}(I_\beta, B_3), \ \_, \_, a_{13}, a_{23}, a_{33}, \ldots, a_{n3}, 0, 0, \ldots, 0) \ \wedge$$

...

$$\text{Val}(\text{In}(I_\beta, B_n), \ \_, \_, \ldots, \_, a_{1n}, a_{2n}, a_{3n}, \ldots, a_{nn}, 0, 0, \ldots, 0)$$

which denotes a series of inputs to the array cells.

(C3)  $\forall 1 \leq i \leq n.$ Val(Mem($M, B_i$), $\square$ $b_i$)

This is equivalent to

$$\text{Val}(\text{Mem}(M, B_1), \square\, b_1) \ \wedge \ \text{Val}(\text{Mem}(M, B_2), \square\, b_2) \ \wedge \ \dots \ \wedge \ \text{Val}(\text{Mem}(M, B_n), \square\, b_n)$$

### (c) Output behavior specification:

Here the dynamic pattern of the *intended* output is specified:

(P1)  $\text{Val}(\text{Out}(O_\alpha, B_n), \bigwedge_{r=0}^{n-1} \_, \bigwedge_{r=1}^{n} (\sum_{i=1}^{n} a_{ri} * b_i), \square\, 0)$

This is equivalent to

$$\text{Val}(\text{Out}(O_\alpha, B_n), \_, \_, \dots, \_, \sum_{i=1}^{n} a_{1i} * b_i, \sum_{i=1}^{n} a_{2i} * b_i, \dots, \sum_{i=1}^{n} a_{ni} * b_i, 0, 0, \dots, 0)$$

The output specification indicates that the output is expected to appear *n* cycles after the first input, followed by a new result every cycle. Total computation time is $2*n$ cycles.

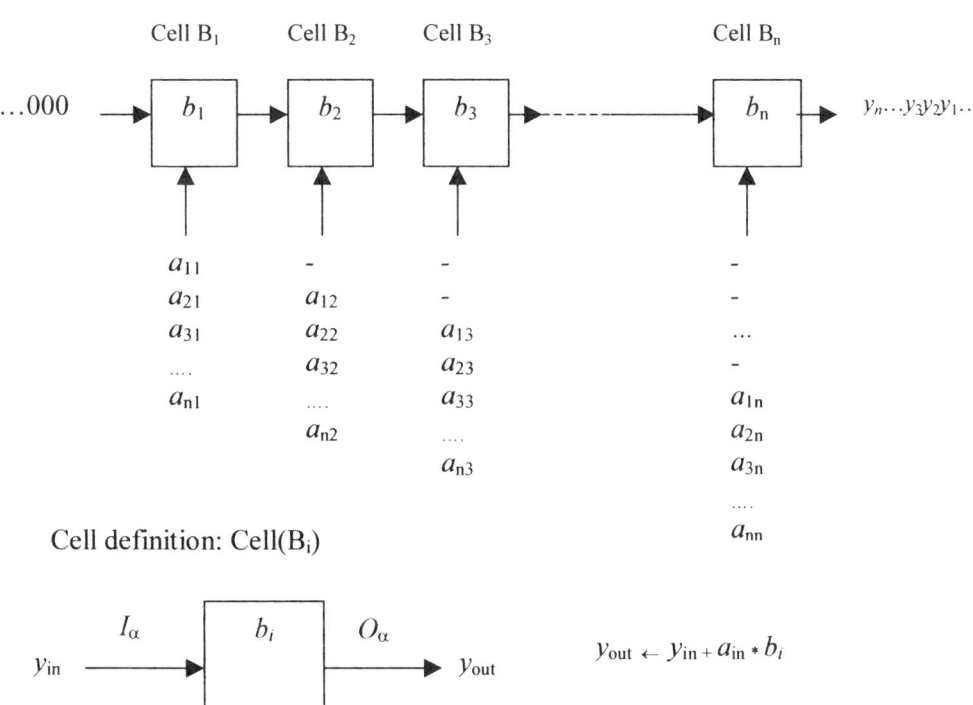

Figure 5.1 1-D systolic array for matrix-vector multiplication and its cell definition

## 5.1.2 Array Verification by STA

The verification requires proving the correctness of

<Specifications (A1) to (A2)> ∧ <Specification (B1)> ∧ <Specifications (C1) to (C3)>

→ <Specification (P1)>

## 5.1.2.1 Method 1: Output Derivation and Comparison

In this approach, the procedure of Section 4.3 is followed. Basically, we use Specifications (A1), (A2), (B1), (C1), (C2), and (C3) as premises, applying STA axioms and rules and conventional logic to derive the output behavior of the array by output propagation. We then compare this to Output Specification (P1) to see if they are equal. If they are, the correctness of the array is verified. These are briefly shown in the following steps:

**Step (S1):** The input behavior of cell $B_1$ is extracted from Specifications (C1) to (C3):

$$\text{Val}(\text{In}(I_\alpha, B_1), \square\ 0) \wedge \text{Val}(\text{In}(I_\beta, B_1), a_{11}, a_{21}, a_{31}, \ldots, a_{n1}, \square\ 0) \wedge \text{Val}(\text{Mem}(M, B_1), \square\ b_1))$$

**Step (S2):** By applying (A1) and the result of (S1) on cell function (B1), the output of cell $B_1$ is:

$$\text{Val}(\text{Out}(O_\alpha, B_1),\ \_, \square\ 0 + (a_{11}, a_{21}, a_{31}, \ldots a_{n1}, \square\ 0) * \square\ b_1)$$

Applying STA Axiom (1) and Rule (1) we get

$$\text{Val}(\text{Out}(O_\alpha, B_1),\ \_, (a_{11}*b_1), (a_{21}*b_1), (a_{31}*b_1), \ldots, (a_{n1}*b_1), \square\ 0)$$

**Step (S3):** Applying wire connection function specification (applied to local connection from cell $B_1$ to cell $B_2$), output of cell $B_1$ (derived at step (S2)) becomes input $I_\alpha$ to cell $B_2$. Together with other inputs to $B_2$, we obtain the input behavior for $B_2$:

$$\text{Val}(\text{In}(I_\alpha, B_2),\ \_, (a_{11}*b_1), (a_{21}*b_1), (a_{31}*b_1), \ldots, (a_{n1}*b_1), \square 0)\ \wedge$$
$$\text{Val}(\text{In}(I_\beta, B_2),\ \_, a_{12}, a_{22}, a_{32}, \ldots, a_{n2}, \square 0)\ \wedge$$
$$\text{Val}(\text{Mem}(M, B_2),\ \square b_2)$$

**Step (S4):** Applying (A1) and the result of (S3) on cell function Specification (B1) we derive the output of cell $B_2$ as:

$$\text{Val}(\text{Out}(O_\alpha, B_2),\ \_, ((\_, (a_{11}*b_1), (a_{21}*b_1), \ldots, (a_{n1}*b_1), \square 0) + (\_, a_{12}, a_{22}, \ldots, a_{n2}, \square 0) * \square b_2))$$

Applying STA Axioms (1) and (2) and Rule (1), we have

$$\text{Val(Out}(O_\alpha, B_2), \_, ((\_, (a_{11}*b_1), (a_{21}*b_1), \dots, (a_{n1}*b_1), \square 0) + (\_, a_{12}*b_2, a_{22}*b_2, \dots, a_{n2}*b_2, \square 0)))$$

Simplifying using STA Axioms (1), (2), and Rule (3), we get

$$\text{Val(Out}(O_\alpha, B_2), \_, \_, (a_{11}*b_1 + a_{12}*b_2), (a_{21}*b_1 + a_{22}*b_2), \dots, (a_{n1}*b_1 + a_{n2}*b_2), \square 0)$$

**Step (S5):** The above process propagates until the output of the last cell is obtained:

$$\text{Val(Out}(O_\alpha, B_n), \_, \_, \dots, (a_{11}*b_1 + a_{12}*b_2 + \dots + a_{1n}*b_n), (a_{21}*b_1 + a_{22}*b_2 + \dots + a_{2n}*b_n), \dots,$$
$$(a_{n1}*b_1 + a_{n2}*b_2 + \dots + a_{nn}*b_n), \square 0)$$

which is

$$\text{Val(Out}(O_\alpha, B_n), \bigwedge_{r=0}^{n-1} \_, \bigwedge_{r=1}^{n} (\sum_{i=1}^{n} a_{ri}*b_i), \square 0)$$

Since this output behavior from (S5) is identical to the output behavior specification (P1), the correctness of the systolic array is thus verified.

## 5.1.2.2 Method 2: Mathematical Induction

The principle of mathematical induction for array verification is applied in this section. Let $P(n)$ be

<Specifications (A1) to (A2)> ∧ <Specification (B1)> ∧ <Specifications (C1) to (C3)>

→ <Specification (P1)>

for the array of size $n$ as shown in Figure 5.1. The output behavior specification (P1) for the same array of a smaller size, say $m$, but with the same input pattern as that of Figure 5.1 for these $m$ cells, is

$$\text{(P1)} \quad \text{Val(Out}(O_\alpha, B_m), \bigwedge_{r=0}^{m-1} \_, \bigwedge_{r=1}^{n} (\sum_{i=1}^{m} a_{ri}*b_i), \square \ 0)$$

Verification steps using mathematical induction are briefly described as follows:

**Step (T1):** *Basis of Induction*: The correctness of $P(1)$ is proved by using steps (S1) and (S2) in the previous method (Subsection 5.1.2.1). The output of cell $B_1$ derived from (S2) is equal to

$$\text{Val(Out}(O_\alpha, B_1), \_, (a_{11}*b_1), (a_{21}*b_1), (a_{31}*b_1), \dots, (a_{n1}*b_1), \square 0)$$

which is

$$\text{Val}(\text{Out}(O_\alpha, B_1), \overset{0}{\underset{r=0}{\bigwedge}} -, \overset{n}{\underset{r=1}{\bigwedge}} (\sum_{i=1}^{1} a_{ri} * b_i), \square 0)$$

which is equal to output specification (P1) with size $m=1$. Hence $P(1)$ is proved.

**Step (T2):** *Inductive Hypothesis*: We assume that $P(k)$ is correct for $1 \le k < n$.

**Step (T3):** *Inductive Step*: If $P(k)$ is correct then the output of cell $B_k$ is given by

$$\text{Val}(\text{Out}(O_\alpha, B_k), \overset{k-1}{\underset{r=0}{\bigwedge}} -, \overset{n}{\underset{r=1}{\bigwedge}} (\sum_{i=1}^{k} a_{ri} * b_i), \square 0)$$

Applying connection function specification to (A2) (applied to local connection from cell $B_k$ to cell $B_{k+1}$), the input $I_\alpha$ to cell $B_{k+1}$ is

$$\text{Val}(\text{In}(I_\alpha, B_{k+1}), \overset{k-1}{\underset{r=0}{\bigwedge}} -, \overset{n}{\underset{r=1}{\bigwedge}} (\sum_{i=1}^{k} a_{ri} * b_i), \square 0)$$

the external input $I_\beta$ to cell $B_{k+1}$, obtained from (C2), is

$$\text{Val}(\text{In}(I_\beta, B_{k+1}), \overset{k-1}{\underset{r=0}{\bigwedge}} -, \overset{n}{\underset{r=1}{\bigwedge}} a_{r,k+1}, \square 0)$$

and memory content of cell $B_{k+1}$, obtained from (C3), is

$$\text{Val}(\text{Mem}(M, B_{k+1}), \square b_{k+1})$$

We derive the input behavior of cell $B_{k+1}$:

$$\text{Val}(\text{In}(I_\alpha, B_{k+1}), \overset{k-1}{\underset{r=0}{\bigwedge}} -, \overset{n}{\underset{r=1}{\bigwedge}} (\sum_{i=1}^{k} a_{ri} * b_i), \square 0) \quad \wedge$$

$$\text{Val}(\text{In}(I_\beta, B_{k+1}), \overset{k-1}{\underset{r=0}{\bigwedge}} -, \overset{n}{\underset{r=1}{\bigwedge}} a_{r,k+1}, \square 0) \quad \wedge$$

$$\text{Val}(\text{Mem}(M, B_{k+1}), \square b_{k+1})$$

Applying (A1) and the above input to cell function (B1) we produce the output behavior of cell $B_{k+1}$

$$\text{Val}(\text{Out}(O_\alpha, B_{k+1}), -, ((\overset{k-1}{\underset{r=0}{\bigwedge}} -, \overset{n}{\underset{r=1}{\bigwedge}} (\sum_{i=1}^{k} a_{ri} * b_i), \square 0) + (\overset{k-1}{\underset{r=0}{\bigwedge}} -, \overset{n}{\underset{r=1}{\bigwedge}} a_{r,k+1}, \square 0) * (\square b_{k+1})))$$

Simplifying using STA Axioms (1) and (2) and Rules (1) and (3) we get

$$\mathrm{Val}(\mathrm{Out}(O_\alpha, B_{k+1}), \bigwedge_{r=0}^{k} -, \bigwedge_{r=1}^{n} ((\sum_{i=1}^{k} a_{ri}*b_i) + a_{r,k+1}*b_{k+1}), \square 0)$$

which is

$$\mathrm{Val}(\mathrm{Out}(O_\alpha, B_{k+1}), \bigwedge_{r=0}^{k} -, \bigwedge_{r=1}^{n} (\sum_{i=1}^{k+1} a_{ri}*b_i), \square 0)$$

Out($O_\alpha$, $B_{k+1}$) is the same as Output Specification (P1) with size $m = k+1$. Hence the correctness of $P(k+1)$ is proved.

By the Principle of Mathematical Induction for systolic array verification, we can therefore conclude that $P(n)$ is correct and the correctness of the array of size $n$ is proved.


## 5.2    A 2-D ARRAY FOR MATRIX-MATRIX MULTIPLICATION

Section 5.1 illustrates how STA can be used to specify and verify a 1-D systolic array; different verification methods and the usage of several constructs have also been shown. In this section, a brief discussion on how STA can be applied to specify and verify a more complicated 2-D systolic array is given. The example adopted here is a 2-D systolic array for matrix-matrix multiplication as shown in Figure 5.2 [KungS85]. Matrix-matrix multiplication operation is common in many image processing and signal processing environments. The matrix-matrix multiplication problem A * B for matrices A and B is defined as follows:

Given $a_{11}, a_{12}, \ldots, a_{1n}, a_{21}, a_{22}, \ldots, a_{nn}$ and $b_{11}, b_{12}, \ldots, b_{1n}, b_{21}, b_{22}, \ldots, b_{nn}$

Compute $y_{11}, y_{12}, \ldots, y_{1n}, y_{21}, y_{22}, \ldots, y_{nn}$

defined by $y_{ij} = \sum_{k=1}^{n} a_{ik}*b_{kj}$

Matrix elements are input in a skewed manner as shown in Figure 5.2. The contents of cell memories $y_{ij}$'s are initialized to 0's. At the end of the computation, the values of the elements of the resulting matrix are the values stored in the cell memories as shown. Computation and data flow are pipelined in two directions.

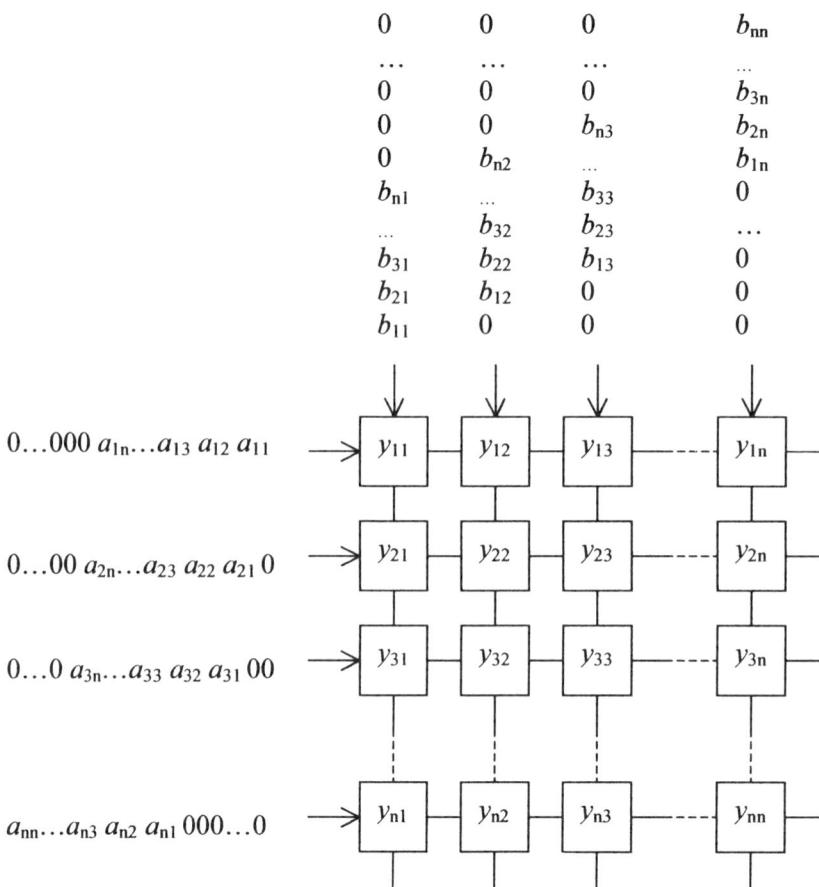

Cell definition:

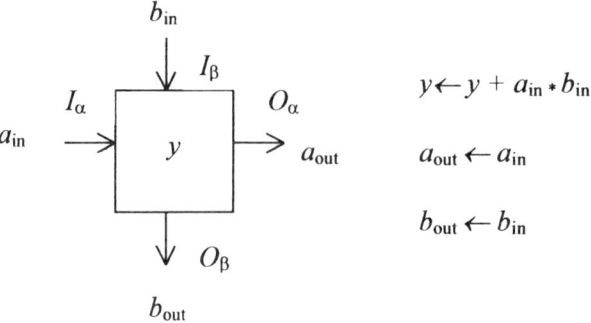

$$y \leftarrow y + a_{in} * b_{in}$$

$$a_{out} \leftarrow a_{in}$$

$$b_{out} \leftarrow b_{in}$$

Figure 5.2 2-D systolic array for matrix-matrix multiplication and its cell definition

## 5.2.1 Array Specification by STA

Let $\mathbf{Y}$ be the array of cells $Y_{11}$, $Y_{12}$ ,..., $Y_{21}$ , $Y_{22}$ ,..., $Y_{nn}$; $\mathbf{Y}_i$ be the vector of cells $Y_{i1}$ , $Y_{i2}$ ,..., $Y_{in}$; and $\mathbf{Y}_j$ be the vector of cells $Y_{1j}$ , $Y_{2j}$ ,..., $Y_{nj}$:

1. **Structure Specification:**

   **(a) Types of components:**

   (A1)  $\models$ Cell($\mathbf{Y}$)     (i.e. $\models$ $\forall 1 \leq i \leq n.\ \forall 1 \leq j \leq n.$ Cell($Y_{ij}$) )

   **(b) Structural connectivity:**

   (A2)  $\models$ $\forall 1 \leq i \leq n.\ \forall 1 \leq j \leq n-1.$ Conn(Out($O_\alpha, Y_{ij}$), In($I_\alpha, Y_{i,j+1}$))

   (A3)  $\models$ $\forall 1 \leq i \leq n-1.\ \forall 1 \leq j \leq n.$ Conn(Out($O_\beta, Y_{ij}$), In($I_\beta, Y_{i+1,j}$))

2. **Behavior Specification:**

   **(a) Functions of individual components and connections:**

   (B1)  The function of the cells used:

   $$\models\ \forall Y_{ij}\ \text{Cell}(Y_{ij}) \wedge \text{Val}(\text{In}(I_\alpha, Y_{ij}), a_{in}) \wedge \text{Val}(\text{In}(I_\beta, Y_{ij}), b_{in}) \wedge \text{Val}(\text{Mem}(M, Y_{ij}), y)$$
   $$\rightarrow\ \text{Val}(\text{Out}(O_\alpha, Y_{ij}), \_, a_{in}) \wedge \text{Val}(\text{Out}(O_\beta, Y_{ij}), \_, b_{in}) \wedge$$
   $$\text{Val}(\text{Mem}(M, Y_{ij}), \_, (y + a_{in} * b_{in}))$$

   Here "*" denotes a multiplication operation.

   The function of a connection (wire connection) is always implied and served as a rule in reasoning process, as before:

   $$\models\ \forall X \forall Y.\ \text{Conn}(X, Y) \wedge \text{Val}(X, x) \wedge \text{Val}(Y, y) \rightarrow \text{Val}(Y, x)$$

   **(b) Input behavior specification:**

   (C1)  $\forall 1 \leq i \leq n.$ Val(In($I_\alpha, Y_{i1}$), $\bigwedge\limits_{r=0}^{i-2} 0, \bigwedge\limits_{r=1}^{n} a_{ir}, \Box 0$)

   This is equivalent to

   Val(In($I_\alpha, Y_{11}$), $a_{11}, a_{12}, ..., a_{1n}, \Box 0$)   and
   Val(In($I_\alpha, Y_{21}$), $0, a_{21}, a_{22}, ..., a_{2n}, \Box 0$)   and ...

which denotes a series of inputs to the array cells from the left (Figure 5.2).

(C2)   $\forall 1 \le j \le n.$  $\mathrm{Val}(\mathrm{In}(I_\beta, Y_{1j}), \bigwedge_{r=0}^{j-2} 0, \bigwedge_{r=1}^{n} b_{rj}, \square\, 0)$

This is equivalent to

$\mathrm{Val}(\mathrm{In}(I_\beta, Y_{11}), b_{11}, b_{21}, ..., b_{n1}, \square 0)$   and
$\mathrm{Val}(\mathrm{In}(I_\beta, Y_{12}), 0, b_{12}, b_{22}, ..., b_{n2}, \square 0)$   and ...

which denotes a series of inputs to the array cells from the top (Figure 5.2).

(C3)   $\forall 1 \le i \le n.$ $\forall 1 \le j \le n.$ $\mathrm{Val}(\mathrm{Mem}(M, Y_{ij}), 0, \square\_\,) \wedge$

$\mathrm{Val}(\mathrm{Out}(O_\alpha, Y_{ij}), 0, \square\_\,) \wedge \mathrm{Val}(\mathrm{Out}(O_\beta, Y_{ij}), 0, \square\_\,)$

This indicates that the memory locations and cell outputs are initialized to 0s.

**(c) Output behavior specification:**

Here the dynamic pattern of the intended output (i.e. the memory content of each cell) is specified:

(P1)   $\forall 1 \le i \le n.$  $\forall 1 \le j \le n.$ $\mathrm{Val}(\mathrm{Mem}(M, Y_{ij}), \bigwedge_{r=0}^{i+j-2} 0, \bigwedge_{r=1}^{n-1} (\sum_{k=1}^{r} a_{ik} * b_{kj}), \square\, (\sum_{k=1}^{n} a_{ik} * b_{kj}))$

That is, the content of the memory location for cell $Y_{ij}$ is 0 for the first $i+j-1$ cycles; then it takes the values $a_{i1}*b_{1j}$, $a_{i1}*b_{1j} + a_{i2}*b_{2j}$, ..., in the cycles thereafter. From cycle $i+j+n-2$ onwards, the memory content of cell $Y_{ij}$ remains as $\sum_{k=1}^{n} a_{ik} * b_{kj}$, which is the value of the element $i,j$ of the resultant matrix as defined earlier. Total computation time is $n+n+n-1$ or $3*n-1$ cycles.

## 5.2.2  Array Verification by STA

The verification requires the proving of the correctness of:

<Specifications (A1) to (A3)> $\wedge$ <Specification (B1)> $\wedge$ <Specifications (C1) to (C3)>

$\rightarrow$ <Specification (P1)>

Output derivation and comparison method (Section 4.3) is used here. The approach is briefly described as follows:

**Step (S1):** Applying (A1) and the input behavior (C1) to (C3) of cell $Y_{11}$ to cell function (B1), the memory content of cell $Y_{11}$ is derived in an iterative manner, using STA axioms and rules, from

$$\text{Val(Mem}(M,Y_{11}),\ 0,((a_{11},a_{12},...,a_{1n},\square 0)*(b_{11},b_{21},...,b_{n1},\square 0) + \text{Val(Mem}(M,Y_{11}))))$$

(5.1)

This is a temporally recursive predicate and the value $\text{Val(Mem}(M, Y_{11}))$ must be updated cycle by cycle.

The initial value ($1^{st}$ cycle) is 0 (C3). Using this as the first value for $\text{Val(Mem}(M, Y_{11}))$, the next ($2^{nd}$) cycle is derived from the above Expression (5.1) using Axiom (1), as $a_{11}*b_{11} + 0$, which is

$$a_{11}*b_{11}$$

Substituting this value for the $2^{nd}$ cycle for $\text{Val(Mem}(M, Y_{11}))$, the $3^{rd}$ cycle is derived from the same Expression (5.1) using Axiom (1), as $a_{12}*b_{21} + a_{11}*b_{11}$, which is

$$a_{11}*b_{11} + a_{12}*b_{21}$$

By continuing this process, the value at the $n+1^{th}$ cycle is

$$a_{11}*b_{11} + a_{12}*b_{21} + ... + a_{1n}*b_{n1}$$

or $\sum_{k=1}^{n} a_{1k}*b_{k1}$. It can be similarly verified that the same value stays for the rest of time, due to continuous 0's at the inputs. Hence

$$\text{Val(Mem}(M,Y_{11}),\ 0,(a_{11}*b_{11}),(a_{11}*b_{11} + a_{12}*b_{21}),...,(\sum_{k=1}^{n} a_{1k}*b_{k1}),\square(\sum_{k=1}^{n} a_{1k}*b_{k1}))$$

which is same as the Specification (P1) with $i=j=1$.

**Step (S2):** The outputs of cell $Y_{11}$ can similar be derived using Specification (B1) (initial values are 0's, as in (C3)), as

$$\text{Val(Out}(O_{\alpha},Y_{11}),\ 0,a_{11},a_{12},...,a_{1n},\square 0)\quad \text{and}$$
$$\text{Val(Out}(O_{\beta},Y_{11}),\ 0,b_{11},b_{21},...,b_{n1},\square 0)$$

**Step (S3):** Again, by connection function these becomes the inputs to cells $Y_{12}$ and $Y_{21}$ respectively; and by cell functional Specification (B1) the memory contents of cells $Y_{12}$ and $Y_{21}$ can similarly be derived as

$$\text{Val(Mem}(M,Y_{12}),\ 0,0,(a_{11}*b_{12}),(a_{11}*b_{12}+a_{12}*b_{22}),...,(\sum_{k=1}^{n}a_{1k}*b_{k2}),\square(\sum_{k=1}^{n}a_{1k}*b_{k2}))$$

and

$$\text{Val(Mem}(M,Y_{21}),\ 0,0,(a_{21}*b_{11}),(a_{21}*b_{11}+a_{22}*b_{21}),...,(\sum_{k=1}^{n}a_{2k}*b_{k1}),\square(\sum_{k=1}^{n}a_{2k}*b_{k1}))$$

which are (P1) with ($i$=1, $j$=2) and ($i$=2, $j$=1) respectively.

**Step (S4):** In the same manner we can derive the outputs of cells $Y_{12}$ and $Y_{21}$ and hence the memory contents of cells $Y_{13}$, $Y_{22}$, and $Y_{31}$. These are again compared to (P1). The process proceeds like a wavefront until the content of the final cell $Y_{nn}$ is derived:

$$\text{Val(Mem}(M,Y_{nn}),\ \bigwedge_{r=0}^{2n-2}0,(a_{n1}*b_{1n}),(a_{n1}*b_{1n}+a_{n2}*b_{2n}),...,(\sum_{k=1}^{n}a_{nk}*b_{kn}),\square(\sum_{k=1}^{n}a_{nk}*b_{kn}))$$

which is (P1) with $i=j=n$.

Hence we have

$$\forall 1\leq i\leq n.\ \ \forall 1\leq j\leq n.$$

$$\text{Val(Mem}(M,Y_{ij}),\ \bigwedge_{r=0}^{i+j-2}0,\bigwedge_{r=1}^{n-1}(\sum_{k=1}^{r}a_{ik}*b_{kj}),\square(\sum_{k=1}^{n}a_{ik}*b_{kj}))$$

which is (P1). The correctness of the array is thus verified.

## 5.3    A BI-DIRECTIONAL ARRAY FOR 1-D CONVOLUTION

One-dimensional convolution is another operation that is very common in many signal processing environments. The 1-D convolution problem is defined as follows:

Given weights $w_1, w_2,..., w_k$ and inputs $x_1, x_2,..., x_n$

Compute $y_1, y_2,..., y_{n-k+1}$

defined by $y_m = \sum_{i=0}^{k-1}w_{i+1}*x_{m+i}$

A bi-directional systolic network for implementing 1-D convolution and its cell definition is given in Figure 5.3 [KungH82]. All $y_m$'s are initialized to zeros and all $w_i$'s are preloaded into the cells.

$x$'s and $y$'s move systolically in opposite directions and consecutive elements of $x$'s and $y$'s are separated by one cycle time. $y_1$ enters the network when $x_1$ has arrived at cell $W_1$. Computation is pipelined and the results $y_m$'s start to appear at the output of cell $W_k$, $2*k-1$ cycles after the first input, followed by a new output every two cycles. Total computation time is $2*n$ cycles.

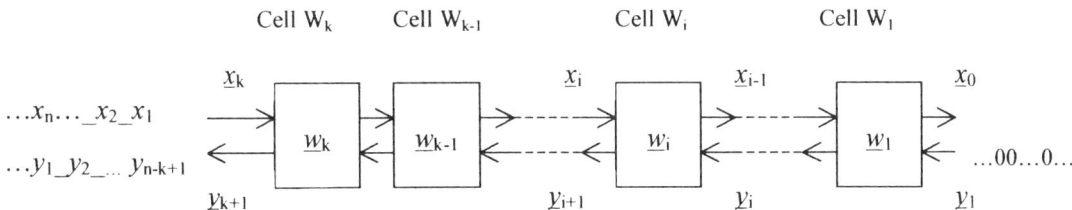

Cell definition:

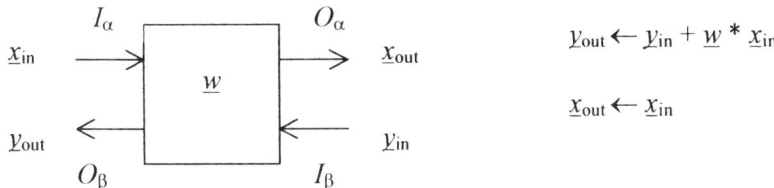

$$y_{out} \leftarrow y_{in} + \underline{w} * \underline{x}_{in}$$

$$\underline{x}_{out} \leftarrow \underline{x}_{in}$$

Figure 5.3 Bi-directional systolic array for 1-D convolution and its cell definition

## 5.3.1 Array Specification by STA

Let **W** denotes the cells $W_1$, $W_2$, ..., $W_k$ collectively.

1. **Structure Specification:**

   (a) **Types of Components:**

   (A1) $\models$ Cell(**W**)          (i.e. $\models$ $\forall 1 \le i \le k$. Cell($W_i$) )

   (b) **Structural Connectivity:**

   (A2) $\models$ $\forall 1 \le i \le k-1$. Conn(In($I_\alpha, W_i$), Out($O_\alpha, W_{i+1}$))

   (A3) $\models$ $\forall 1 \le i \le k-1$. Conn(Out($O_\beta, W_i$), In($I_\beta, W_{i+1}$))

## 2. Behavior Specification:

### (a) Functions of components and interconnections:

The function of the cells used:

(B1) $\models$ $\forall W_i$ $\mathrm{Cell}(W_i) \wedge \mathrm{Val}(\mathrm{In}(I_\alpha, W_i), x_{in}) \wedge \mathrm{Val}(\mathrm{In}(I_\beta, W_i), y_{in}) \wedge \mathrm{Val}(\mathrm{Mem}(M, W_i), w_i)$
$\rightarrow \mathrm{Val}(\mathrm{Out}(O_\alpha, W_i), \_, x_{in}) \wedge \mathrm{Val}(\mathrm{Out}(O_\beta, W_i), \_, (y_{in} + w_i * x_{in}))$

The function of a connection (wire connection) is always implied and served as a rule in reasoning process, as before:

$\models$ $\forall X \forall Y.$ $\mathrm{Conn}(X, Y) \wedge \mathrm{Val}(X, x) \wedge \mathrm{Val}(Y, y) \rightarrow \mathrm{Val}(Y, x)$

### (b) Input behavior specification:

(C1) $\mathrm{Val}(\mathrm{In}(I_\alpha, W_k), \overset{2n-2}{\underset{r=0 \text{ by } 2}{\bigwedge}} (x_{\frac{r+2}{2}}, \_), \square\_)$

This is equivalent to

$\mathrm{Val}(\mathrm{In}(I_\alpha, W_k), x_1, \_, x_2, \_, \dots, x_n, \_, \square\_)$

(C2) $\mathrm{Val}(\mathrm{In}(I_\beta, W_1), \overset{k-2}{\underset{r=0}{\bigwedge}}\_, \overset{2(n-k)}{\underset{r=0 \text{ by } 2}{\bigwedge}} (0, \_), \square\_)$

(C3) $\forall 1 \leq i \leq k.$ $\mathrm{Val}(\mathrm{Mem}(M, W_i), \square w_i)$

### (c) Output behavior specification:

Here the dynamic pattern of the intended output is specified:

(P1) $\mathrm{Val}(\mathrm{Out}(O_\beta, W_k), \overset{2k-2}{\underset{r=0}{\bigwedge}}\_, \overset{2(n-k)}{\underset{r=0 \text{ by } 2}{\bigwedge}} (y_{\frac{r}{2}+1}, \_), \square\_)$

where $y_m = \sum_{i=0}^{k-1} w_{i+1} * x_{m+i}$

This is equivalent to

$\mathrm{Val}(\mathrm{Out}(O_\beta, W_k), \overset{2k-2}{\underset{r=0}{\bigwedge}}\_, y_1, \_, y_2, \_, \dots, y_{n-k+1}, \_, \square\_)$

The output specification indicates that the output is expected to appear $2k\text{-}1$ cycles after the first input, followed by a new result every two cycles. The last result will appear $2n\text{-}1$ cycles after the first input, giving the total computation time of $2n$ cycles.

## 5.3.2  Array Verification by STA

The verification requires the proving of the correctness of:

<Specifications (A1) to (A3)> $\wedge$ <Specification (B1)> $\wedge$ <Specifications (C1) to (C3)>

$\rightarrow$ <Specification (P1)>

The method by solving STA difference equations (Section 4.5) is used here. Unlike previous problems, to solve this problem we express both normal and temporal variables explicitly. Temporal variables (underlined) are used to denote the sequences of values that appear on the input/output links of the cells, as shown in Figure 5.3. Cell $W_1$ denotes the rightmost cell and cell $W_k$ denotes the leftmost cell of the systolic network. Let temporal variables $\underline{x}_i$, $\underline{x}_{i-1}$, $\underline{y}_i$, $\underline{y}_{i+1}$ and $\underline{w}_i$ denote the sequences of values on input/output/memory $I_\alpha, O_\alpha, I_\beta, O_\beta$, and $M$ respectively, for cell $W_i$. Temporal variables $\underline{x}_k$, $\underline{x}_0$, $\underline{y}_{k+1}$, and $\underline{y}_1$ thus denote the sequences of values on array inputs and outputs. Verification is performed in the following steps:

**Step (S1):** The input/output relationship for the *cells* can first be obtained from functional specification (B1) (since (A1)) and expressed using *STA difference equations*:

$$\underline{x}_{i-1} = \_,\underline{x}_i \qquad \text{for } i = 1,2,...,k \tag{S1.1}$$

$$\underline{y}_{i+1} = \_,(\underline{y}_i + \underline{w}_i * \underline{x}_i) \qquad \text{for } i = 1,2,...,k \tag{S1.2}$$

Note that to obtain the above expressions, connectivity specifications (A2), (A3), and wire connection function are used, although in a trivial manner.

**Step (S2):** We derive the *input/output relation* for the *array*:

Applying Theorem 1 (Chapter 3), the solution of Equation (S1.1) is

$$\underline{x}_i = \bigwedge_{r=0}^{k-i-1} \_,\underline{x}_k \qquad \text{for } i = 0,1,2,...,k \tag{S2.1}$$

Substitute the value of $\underline{x}_i$ given by Equation (S2.1) into Equation (S1.2) gives

$$\underline{y}_{i+1} = \_,(\underline{y}_i + \underline{w}_i * (\bigwedge_{r=0}^{k-i-1} \_, \underline{x}_k))$$

$$= (\_,\underline{y}_i) + (\_,\underline{w}_i * (\bigwedge_{r=0}^{k-i-1} \_, \underline{x}_k))$$

$$= (\_,\underline{y}_i) + (\bigwedge_{r=0}^{k-i} \_, \underline{w}_i * \underline{x}_k)$$

The above manipulations are performed using STA Axioms (1) and (2), Rules (1) and (3), as well as the fact that $\underline{w}_i = \square w_i$.

Hence $y_{i+1}$ is summarized by the equation

$$\underline{y}_{i+1} = (\_,\underline{y}_i) + (\bigwedge_{r=0}^{k-i} \_, \underline{w}_i * \underline{x}_k) \qquad \text{for } i = 1,2,...,k \tag{S2.2}$$

which is the form specified in Equation (T2.1) of Theorem 2 (Chapter 3). Hence, by Theorem 2, the solution is

$$\underline{y}_t = (\bigwedge_{r=0}^{t-2} \_, \underline{y}_1) + \sum_{j=1}^{t-1}(\bigwedge_{r=0}^{j-2} \_,(\bigwedge_{r=0}^{k-t+j} \_, \underline{w}_{t-j} * \underline{x}_k)) \qquad \text{for } t = 2,3,...,k+1$$

Hence for $t = k+1$ we have

$$\underline{y}_{k+1} = (\bigwedge_{r=0}^{k-1} \_, \underline{y}_1) + \sum_{j=1}^{k}(\bigwedge_{r=0}^{j-2} \_,(\bigwedge_{r=0}^{j-1} \_, \underline{w}_{k-j+1} * \underline{x}_k))$$

$$= (\bigwedge_{r=0}^{k-1} \_, \underline{y}_1) + \sum_{j=1}^{k}(\bigwedge_{r=0}^{2j-2} \_, \underline{w}_{k-j+1} * \underline{x}_k)$$

The above manipulation is performed using STA Rule (2). Hence $y_{k+1}$ is summarized by

$$\underline{y}_{k+1} = (\bigwedge_{r=0}^{k-1} \_, \underline{y}_1) + \sum_{j=1}^{k}(\bigwedge_{r=0}^{2j-2} \_, \underline{w}_{k-j+1} * \underline{x}_k) \tag{S2.3}$$

**Step (S3):** We now need to see if this is the same as Output Specification (P1). (P1) is given by:

(P1)  $\text{Val}(\text{Out}(O_\beta, W_k), \bigwedge_{r=0}^{2k-2} \_, \bigwedge_{r=0 \text{ by } 2}^{2(n-k)} (y_{\frac{r}{2}+1}, \_), \square\_)$

where $y_m = \sum_{i=0}^{k-1} w_{i+1} * x_{m+i}$

We substitute each input temporal variable in the I/O relation (Equation (S2.3)) by the corresponding normal variables and constants to obtain the output behavior of the array. From input specifications (C1)-(C3), we have

$$\underline{x}_k = \bigwedge_{r=0 \text{ by } 2}^{2n-2} (x_{r+2}, \_), \Box \frac{\Box}{2}$$

$$\underline{y}_1 = \bigwedge_{r=0}^{k-2} \_, \bigwedge_{r=0 \text{ by } 2}^{2(n-k)} (0, \_), \Box \_$$

$$\forall 1 \le i \le k. \quad \underline{w}_i = \Box w_i$$

Substitute these into the array I/O relation (Equation (S2.3)) obtained from Step (S2), we have

$$\underline{y}_{k+1} = (\bigwedge_{r=0}^{k-1} \_, \bigwedge_{r=0}^{k-2} \_, \bigwedge_{r=0 \text{ by } 2}^{2(n-k)} (0, \_), \Box \_) + \sum_{j=1}^{k} (\bigwedge_{r=0}^{2j-2} \_, (\Box w_{k-j+1}) * (\bigwedge_{r=0 \text{ by } 2}^{2n-2} (x_{r+2}, \_), \Box \frac{}{2} \_))$$

**Step (S4):** Applying STA Axioms (1), (2), and Rules (1), (2), we derive

$$\underline{y}_{k+1} = (\bigwedge_{r=0}^{2k-2} \_, \bigwedge_{r=0 \text{ by } 2}^{2(n-k)} (0, \_), \Box \_) + \sum_{j=1}^{k} (\bigwedge_{r=0}^{2j-2} \_, \bigwedge_{r=0 \text{ by } 2}^{2n-2} (w_{k-j+1} * x_{r+2}, \_), \Box \frac{}{2} \_)$$

Applying Axioms (1), (2), and Rules (2), (3), we have

$$\underline{y}_{k+1} = (\bigwedge_{r=0}^{2k-2} \_, \bigwedge_{r=0 \text{ by } 2}^{2(n-k)} (0, \_), \Box \_) + (\bigwedge_{r=0}^{2k-2} \_, \bigwedge_{r=0 \text{ by } 2}^{2(n-k)} (\sum_{j=1}^{k} (w_j * x_{j+\frac{r}{2}}), \_), \Box \_)$$

Further applying STA Axioms (1) and (2) we derive

$$\underline{y}_{k+1} = \bigwedge_{r=0}^{2k-2} \_, \bigwedge_{r=0 \text{ by } 2}^{2(n-k)} (\sum_{j=1}^{k} (w_j * x_{j+\frac{r}{2}}), \_), \Box \_$$

$$= \bigwedge_{r=0}^{2k-2} \_, \bigwedge_{r=0 \text{ by } 2}^{2(n-k)} (y_{\frac{r}{2}+1}, \_), \Box \_$$

where $y_m = \sum_{i=0}^{k-1} w_{i+1} * x_{m+i}$

Hence the output behavior of the array is

$$\text{Val}(\text{Out}(O_\beta, W_k), \bigwedge_{r=0}^{2k-2} \_, \bigwedge_{r=0 \text{ by } 2}^{2(n-k)} (y_{\frac{r}{2}+1}, \_), \Box \_)$$

$$\text{where } y_m = \sum_{i=0}^{k-1} w_{i+1} * x_{m+i}$$

**Step (S5):** The above expression is already in the canonical form used and it is equal to the output specification (P1). The correctness of the array is thus verified.

$\Box$

# Chapter 6

# VSTA: A SPECIAL PURPOSE FORMAL VERIFIER FOR SYSTOLIC DESIGNS

## 6.1   INTRODUCTION

In verifying the design correctness of a specific class of architectures, special purpose formal design verifier has the advantage of being able to exploit the attributes of that architecture class to improve the efficiency in the design verification process. Such development is important due to the fact that architecture design verification using general purpose theorem prover is usually extremely time consuming. This chapter briefly presents a Prolog-based verifier, VSTA, that we developed to automate formal design verification process for systolic array architectures.

We use Prolog to help automate our techniques due to Prolog's powerful pattern matching and automatic back-tracking mechanisms, its popularity and quality, its similarity in representing facts with STA, and its wide acceptance in lower level module and circuit verification [Maru85, Fuji83] (so as to achieve multilevel reasoning). In the next chapter and in the appendix, we describe the application of our tool to verify the correctness of two systolic array designs. Executing the verifier on our workstation shows that a typical array design can usually be verified in less than 10 minutes.

Very few of the techniques mentioned in the literature survey in Chapter 2 are developed into automated tools for systolic design verification. An example is Purushothaman and Subrahmanyam's mechanical certification and the use of Boyer-Moore theorem prover [Puru89], and a few others [Hoare92]. In this chapter, we extend our verification technique and present a suitable Prolog-based verifier VSTA we developed to automate design verification [Ling93, Shih95]. Attributes of systolic arrays are exploited in these techniques as well as in the verifier to improve efficiency in design verifications. Our verifier is designed with the following goals in mind:

1. Ease of encoding, debugging, and manipulating the steps of execution;
2. Quality and correctness of the verification process; and
3. Fast execution time.

We have applied our tool to verify several systolic arrays designed for DFT, QR decomposition, LU decomposition, matrix multiplication, and convolution algorithms.

## 6.2    INDUCTIVE DESIGN VERIFICATION TECHNIQUES

Most hardware systems have irregular structures and as such the concept of induction cannot be applied to verify their correctness. However, due to the repeatability, regularity, and locality nature of systolic arrays, we found that induction techniques are very suitable for the formal verification of their designs. The principle exploits these properties to construct a proof that an array of any size is correct. The strategy is especially efficient for large arrays or arrays of parameterized sizes, since the number of steps in the procedure does not depend on the number of cells in general. Our verifier VSTA adopts the following four induction techniques:

1. Regular Mathematical Induction
2. Structured Induction
3. Double Induction
4. Reverse Induction

These techniques are described in detail in Chapter 4. Our verifier developed to automate these induction techniques with the use of backward chaining is discussed in the next section. In Chapter 7, we briefly illustrate how different induction techniques can be utilized to verify a triangular array for LU decomposition.

## 6.3    VSTA AUTOMATION TECHNIQUE

We present briefly, in this section, a formal verifier software developed for inductive proof techniques discussed in earlier chapters. A popular logic programming language, Prolog, is used to implement the verifier. We adopt Prolog to automate STA verification for systolic arrays due to the following reasons:

1. Its usefulness and similarity to STA in representing bodies of facts in predicate forms (hence, efficiency in encoding, understanding, and debugging).
2. Its power in symbolic manipulation: Prolog's pattern matching and automatic backtracking mechanisms are very useful in implementing logical inferences.
3. Its ability to complete a proof within a reasonable amount of time and its implementation minimizes the unnecessary lower abstraction layers.
4. Its popularity and its wide acceptance for lower level module and circuit specifications and verifications. This allows the forming of a multilevel reasoning system.

In our verifier, called VSTA, temporal variables are encoded using Prolog list structure. Slight dissimilarity between Prolog and STA is bridged by a few operator definitions in Prolog. The representation of STA axioms, rules, and theorems are expressions in Prolog. They can be decomposed and built up by Prolog predicates. These are represented as abstract objects in VSTA. Such meta-linguistic abstraction can be manipulated easily by our verifier (which is a metainterpreter). The verification process is further described in the form of a proof tree later in this chapter.

Systolic design description is input to the verifier in three different forms (clauses). If both the antecedents and consequent exist in a logic implication, the resulting clause is an "inference rule". Component function specification falls into this category. If a clause has only the consequent part, the clause is a "fact". Structural specification and input behavior specification fall into this category. On the other side, if a clause has only the antecedent part, the clause is a "question". The output specification to be proved is provided by the user as a question to the verifier. The input to our verifier thus consists of (1) declaring constrained facts (quantified predicate-type specifications), (2) defining inference rules (implication-type specifications), and (3) asking questions (a yes/no question on output specification). Our Prolog-based verifier is a man-machine interactive tool using induction, backward chaining, and rewriting to perform a proof of the goal. When temporal variables are not normalized, normalization techniques are applied. The output specification is formatted as a yes/no question input to the verifier and is treated as the goal to be proved.

Our verifier takes care of different kinds of goals. Backwards proof takes place by matching a goal with the consequent of a rule; the antecedents become the subgoals. This unfolding process (replacing the consequent part by the corresponding antecedents) repeats until sufficient number of facts are matched, in which case the proof of that goal is completed. Since STA specification involves some bound quantifier (e.g. $\forall i \ 1 \leq i \leq n$), a fact (or the antecedents of a rule) may have constraints. If the constraints are satisfied, the fact is valid and thus it can be used in a deduction step.

VSTA uses backward chaining to implement induction. The goal of proving $P(n)$ is set as a yes/no question and is divided into proving two subgoals: proving $P(n_0)$ (or $P(N)$) and $P(k+1)$ (or $P(k-1)$), assumed $P(k)$. Each subgoal is then treated as goal with the corresponding antecedents treated as subgoals to be proved. This process is done recursively, matching goals against consequents, setting up antecedents as subgoals, and backtracking in cases of failure. If all the subgoals can be satisfied or validated, the goal is proved and the answer to the yes/no question will be yes. For the purpose of illustration, a simple example of this is depicted as a proof tree shown in Figure 6.2, for verifying a 1-D systolic array for matrix-vector multiplication (Figure 6.1, also described in Chapter 5). At certain points in the tree, the subgoals are simply the constrained facts (i.e. structural and input specifications themselves), and are therefore satisfied, or auxiliary predicates, which are validated. These are shown as leaf nodes in the tree. Implication specifications such as the component functional specifications can be used as rules of inference to help set up subgoals. Our verifier allows the user to select appropriate rules in a deduction step to improve efficiency. The fact that Prolog representation is close to our STA notation in representing bodies of facts in predicate forms makes user control and debugging easy.

Referring to the proof tree of Figure 6.2, the root of the proof tree is our goal (i.e. output specification). Proof of this goal consists of proving the base and the induction step (the two branches from the root). The base proof is a constrained goal consisting of constraints (left branch) and a goal body. The constraints (the size of the base case) is a conjunction of subgoals which checks the number of cells and the number of input cycles. These subgoals are satisfied due to the structural specification and the input specification applied to the base case. The goal body of the base proof (its right branch) is the output "Val" specification part of $P(n_0)$. This value predicate is the output specification in STA for the array of size $n_0$. The verifier matches this predicate with a consequent in the implication of the component function specification ((B1) in the Figure). This implication is treated as an inference rule and the verifier then unfolds this functional specification, replacing the consequent with the corresponding antecedents (with proper substitution) as subgoals to be proved. This is called the functional specification goal, which is again, a constrained goal. The constraint subgoal is similarly satisfied by the quantification aspect of the component type specification. The goal body of the constrained goal is a conjunction goal of four elements: one component type goal and three input specification goals. These four elements are constrained facts given in the specifications, and are therefore satisfied. Hence the base proof is completed.

The induction step proof (proof of $P(k+1)$) is done in the same manner except that in one of the input specification goal we have another unfold. This unfold is due to the fact that the input to the level of cells forming the array of size $k+1$ depends on the output of the array of size $k$. This is done by unfolding the function specification of an interconnection, which is simply to transmit signals from one end to another without alteration. This is given as (B2) in the Figure. Function specification is applied to the interconnections, which connect the array of size $k$ to cells to form the array of size $k+1$. The antecedents of this implication requires the proof of the existence of appropriate interconnections (connection goal on its left branch) and correct output value from the array of size $k$ (output specification goal for size $k$ on the right branch). The connection goal is satisfied by structural connectivity specifications while the output specification goal is validated by the inductive hypothesis (correctness of $P(k)$). In summary, the proof tree shows the decomposition of the proof procedure down to the leaf level. Each formula at the leaf level is either a constrained fact or a goal that can be validated by auxiliary predicates. Since all the leaf goals are satisfied (constrained facts) or valid (auxiliary predicates), the proof is completed. Our Prolog verifier uses depth-first search in the proof process.

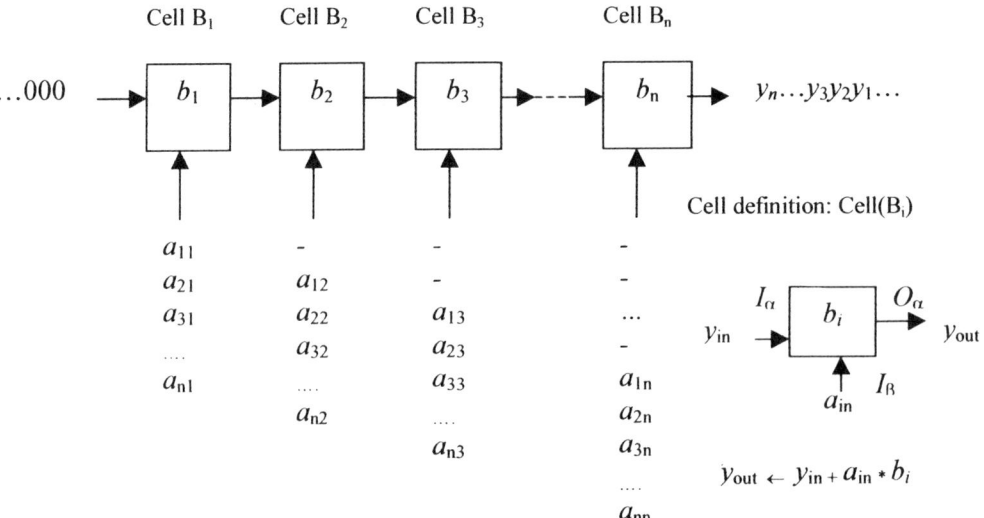

Figure 6.1 1-D systolic array for matrix-vector multiplication and its cell definition

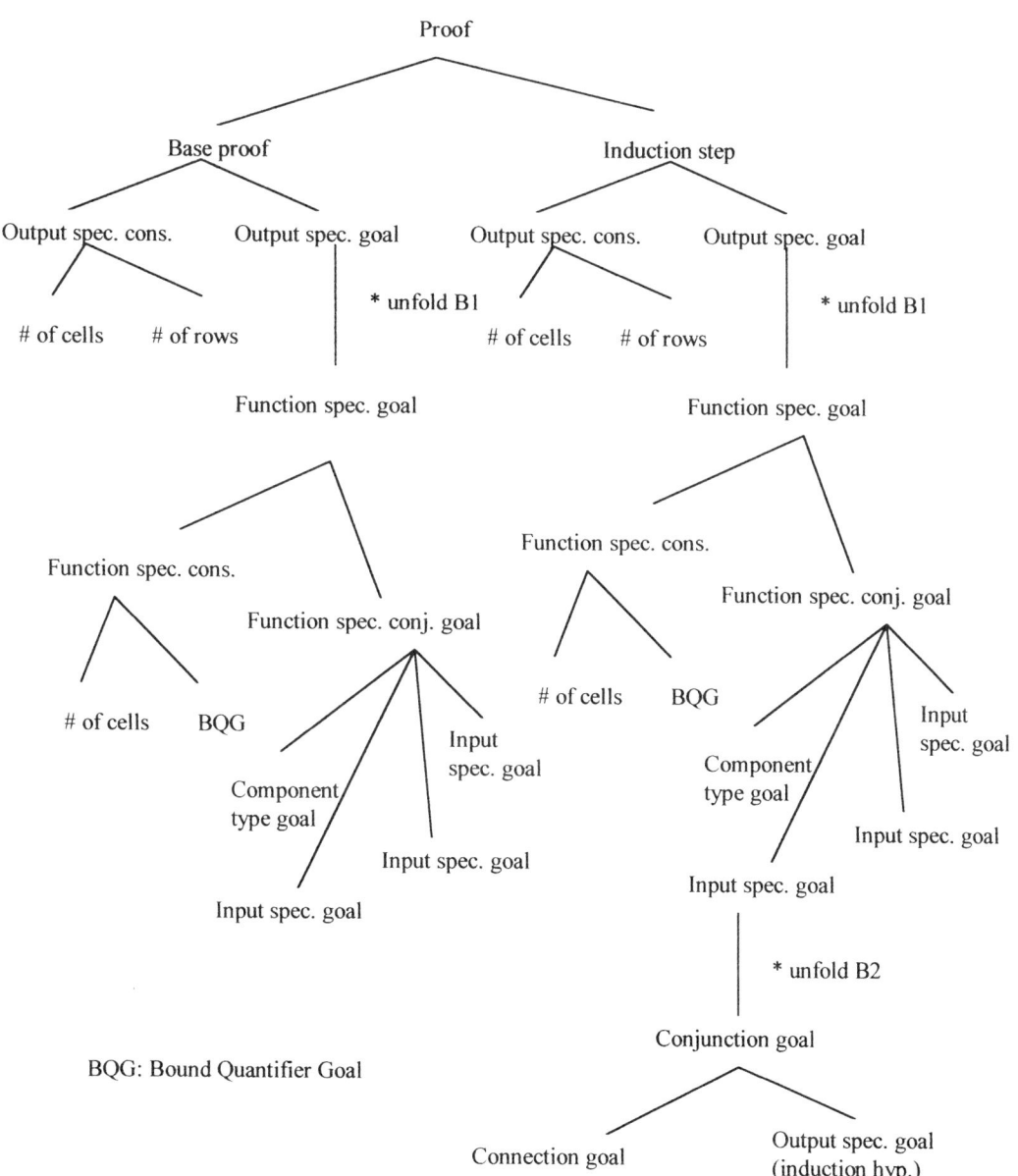

Figure 6.2 A proof tree showing the verification process for the
1-D matrix-vector multiplication systolic array

## 6.4    VSTA VERIFIER STRUCTURE

An overview of the structure of our verifier VSTA is given in Figure 6.3. The user provides three kinds of information to VSTA:

- the specifications, which describe the architecture,
- the strategy decision, which is interactively constructed and selects an appropriate rule to apply in an unfolding step, and
- the architecture specific heuristics, which suggest temporal variable values used in the unfolding process.

VSTA consists of several sub-processors. Each sub-processor handles a specific task. For instance, an induction engine generates the base goal and the induction step goal and passes them to the main processor, which has a central control routine and three databases holding STA specifications for the architecture, STA rules, and STA axioms. The central control routine picks a goal (the first goal in the goal list to be solved), passes it to one of the sub-processors, which either solves the goal (in the cases of the three testers), or transforms the goal to a new goal (in the cases of the temporal value normalizer and the unfolding rule solver). The three testers receive information from the STA specification database and the STA axiom database and send messages back to the main processor to indicate whether a goal is solved. The temporal value normalizer normalizes a temporal value if it is not normalized; this process transforms the existing goal to a new goal, which is passed to the main processor. The normalization process invokes the STA rule database and the STA axiom database. Finally, the unfolding rule solver looks at the domain specific heuristic rule database (which is constructed based on user's suggestion) and unfolds the conclusion (consequent) of a rule to the hypotheses (antecedents) of the rule. The new goal (the hypotheses) is then passed back to the main processor of VSTA. All of these verification processes are executed semi-automatically. At some steps the user is asked to give decisions on using a specification or a rule (or an axiom) and to provide domain specific heuristics in the process.

A sample user interface and a sample session of VSTA verification process are provided in Appendix A.

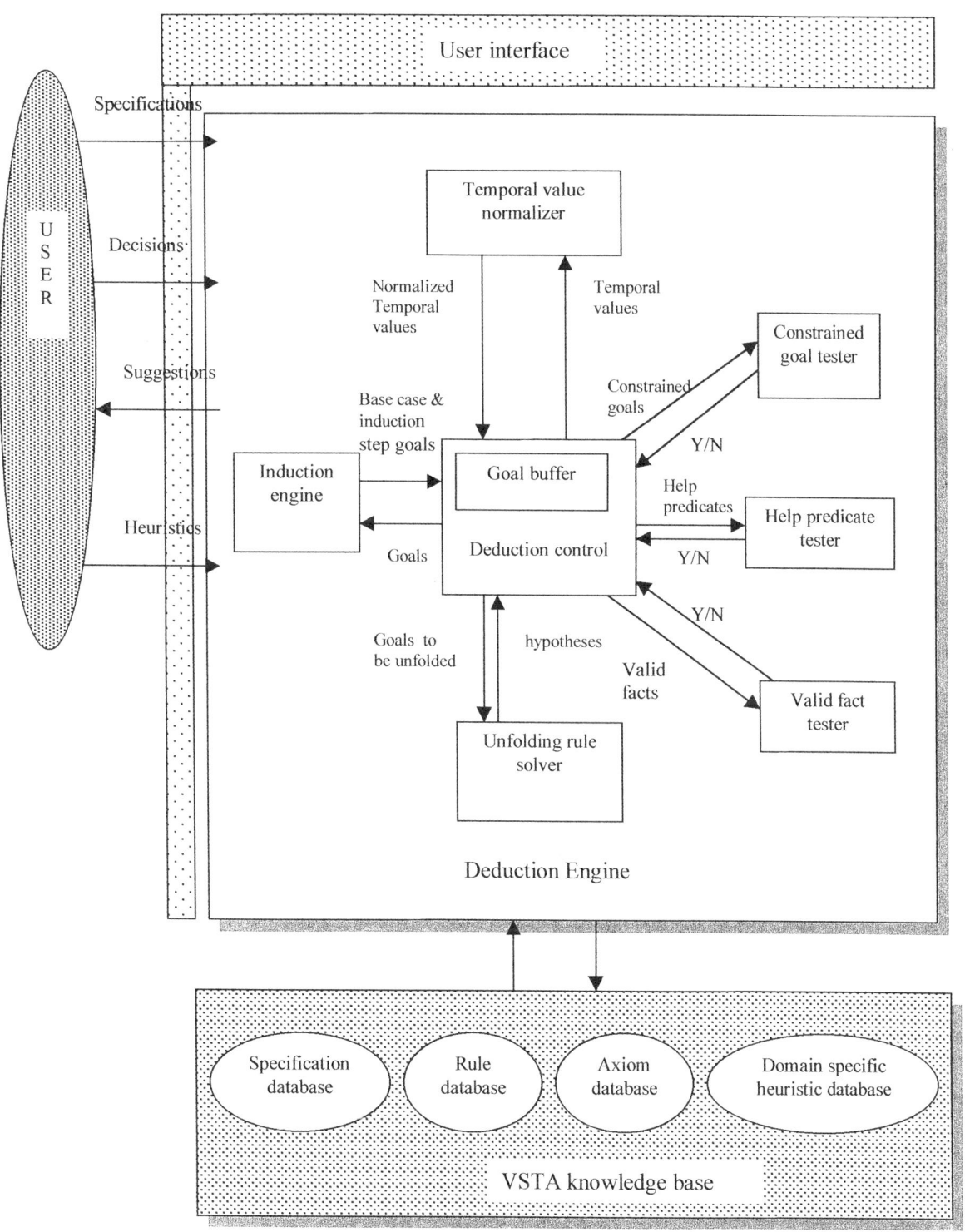

Figure 6.3 Structure of verifier VSTA

# Chapter 7

# VERIFYING THE CORRECTNESS OF A SYSTOLIC ARRAY FOR LU DECOMPOSITION

## 7.1 A SYSTOLIC ARRAY FOR LU DECOMPOSITION

A well-known technique for solving linear equations is the LU decomposition factoring process. The procedure is essentially that of factoring a given matrix C into a lower triangular matrix L and an upper triangular matrix U, where C=L*U. In this chapter, we briefly discuss how the techniques presented are used to prove the correctness of a triangular systolic array designed for LU decomposition. The LU decomposition problem and its corresponding systolic array design are described in several articles (e.g. [KungS88]). A systolic array designed for LU decomposition is shown in Figure 7.1 [KungS88], which is a rather complex example. We aim to provide a formal specification of the design and to verify this design by formal verification using STA [Shih95, Ling95]. The process also illustrates how a proof can be conducted by VSTA [Ling93, Shih95]. Matrix C to be decomposed is fed into the top of the array shown in the Figure. The lower matrix (we call it matrix A) is output from the $M$ cells while the upper matrix (matrix B) is stored in the registers in the $M$ and $N$ cells. Each cell has three internal registers for storing intermediate values or final results, as shown in the cell definitions (Figure 7.1). The elements of input matrix C is modified and passed to the next level of cells in the vertical direction in each iteration. Matrix B stored in the registers is also updated step by step toward its final value. The elements of matrix A are obtained from matrix C initially and propagated one cell after each delay cycle to the next column of the array in the horizontal direction. The operation of each cell is defined in the Figure.

Proving the correctness of the array is not trivial due to its complicated data flow and operation. The structural and behavioral extensions in the $i$ direction going from a size $k$ array to a size $k+1$ array are different from those in the $j$ direction. A regular double induction technique cannot be applied here in a straightforward manner. Besides topology, data flow is also less homogeneous compared to many other arrays. Moreover, the array consists of two different types of cells, namely, $N_{ij}$'s and $M_i$'s.

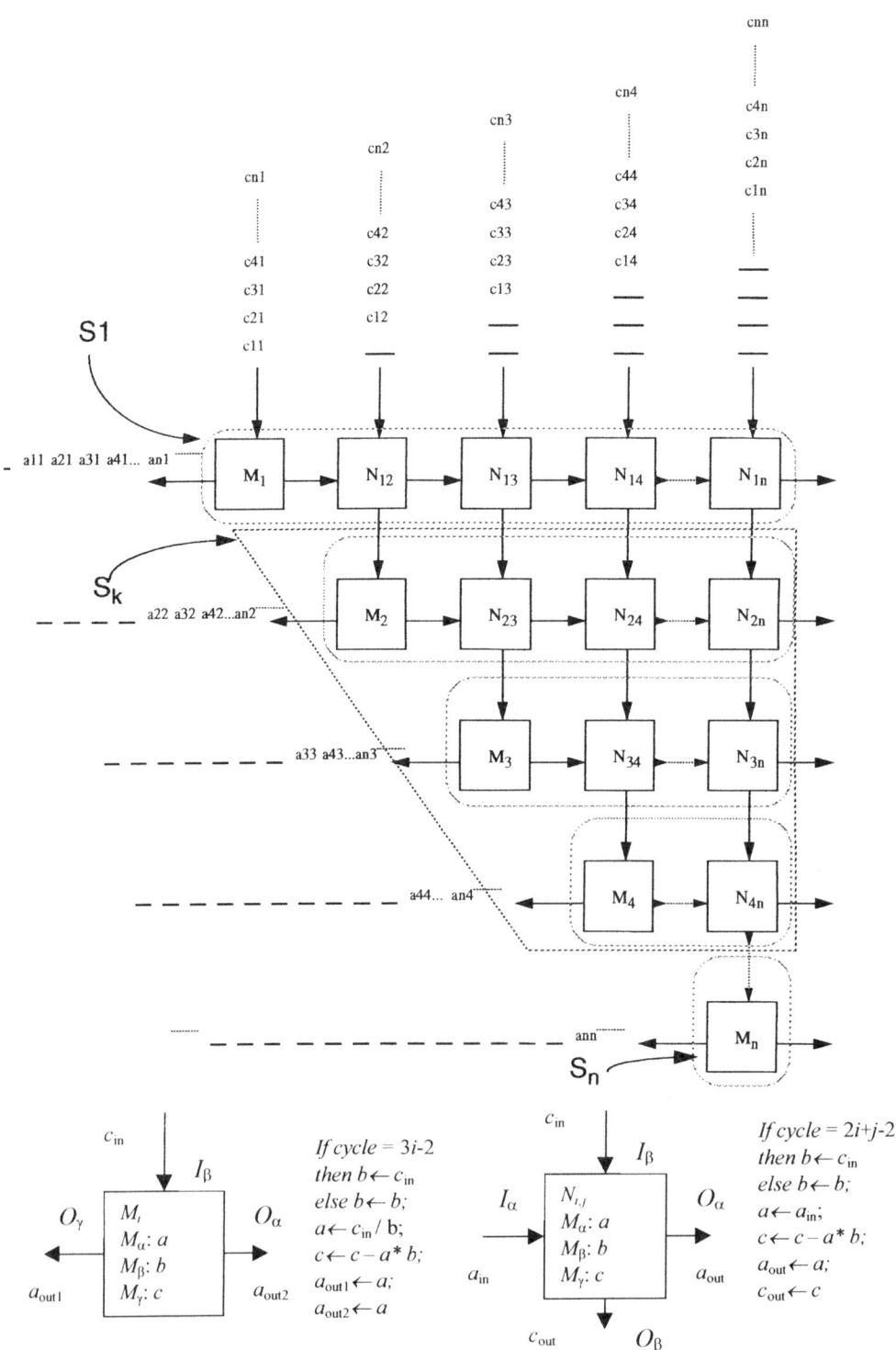

Figure 7.1 A triangular systolic array for LU decomposition and its cell definition

To prove the correctness of the LU decomposition array of Figure 7.1, we decompose the array into three subarrays $S_1$, $S_k$, and $S_n$, as shown in Figure 7.1. We need to prove these three different subarrays. $S_1$ is considered as a separate case due to different inputs; and $S_n$ is another separate case since it does not consist of cells $N_{ij}$ and the corresponding data flow. The 1-D array $S_1$ can be proved easily by regular mathematical induction given the inputs, the structural, and the cell functional specifications (in a similar manner as that of Figure 5.1). Based on the correctness of $S_1$ we can confirm the outputs from this $S_1$ sub-array, and hence the inputs to the sub-array $S_k$. With the correct inputs and the given specifications, reverse induction is now used to prove the array $S_k$, because this array is trapezoidal, forming a part of the triangular array as shown in Figure 7.1, with the longest row closest to the inputs. We should therefore prove this longest row first and use it as the base for reverse induction. Finally the sub-array $S_n$ can be proved by trivial rewriting (simple deduction) technique since only one cell is involved. This completes the proof. An abstract proof tree (consists of several sub-proof trees) for the LU decomposition example is given in Figure 7.2.

## 7.2 FORMAL SPECIFICATION OF THE ARRAY

The LU decomposition systolic array is specified by the specifications and definitions in this section. Specifications (C1) and (C2) give the symbolic temporal value of the input matrix, which is fed into the array from the top. Matrix C is decomposed into matrix A (the lower matrix) and matrix B (the upper matrix) after $3N-2$ cycles, where $N$ is the width (or height) of the input matrix, or size of the array. Matrix A is output from the port $O_\gamma$ of the $M$ cells (specified in specification (P1)) while matrix B is stored in registers $M_\beta$ of cells $M$'s and $N$'s (indicated by specifications (P2) and (P3)). Specifications (A1) and (A2) declare the types of cells in the array. Specifications (A3) to (A6) describe the connection between cells. Behavior specification (B1) indicates that the temporal value is passed from a port to another (wire connection). Specification (B2) says that if the input to the cell $M_i$ from port $I_\beta$ is $\underline{c_{input}}$, then the three register contents $M_\alpha$, $M_\beta$, and $M_\gamma$, and the outputs $O_\alpha$ and $O_\gamma$ will have their values as indicated in (B2). Specification (B3) is similar to Specification (B2) except that the functional specification of cell $N_{i,j}$ is described. Note that temporal values are underlined.

According to the algorithm of LU decomposition, the LU Decomposition Equality Definitions are given. They are used in VSTA as domain specific heuristics. Intermediate specifications (T1), (T2), and (T3) are also added since proving (P1), (P2), and (P3) needs these three specifications. Thus, (T1), (T2), and (T3) are proved first and stored in the axiom database before proof (P1), (P2), or (P3) is proceeded. The specification of LU decomposition systolic array is shown in this section. Section 7.3 shows the proofs for specifications (T1), (T2), (T3), (P1), (P2), and (P3).

In the specification below, a new construct "$\downarrow$" is introduced; $\underline{t}\downarrow_i$ extracts the $i^{th}$ normal variable from the temporal variable $\underline{t}$:

$$\underline{t}\downarrow_i \equiv_{def} i^{th} \text{ normal variable of } \underline{t}$$

## 1.  Structure Specification:

### (a) Types of components:

(A1)  $\forall i \quad 1 \le i \le N. \quad I_{PM}(M_i)$

(A2)  $\forall i \; 1 \le i \le N-1, \forall j \; i+1 \le j \le N. \quad I_{PN}(N_{i,j})$

Here $I_{PM}$ and $I_{PN}$ denote two different types of cells.

### (b) Structural connectivity:

(A3)  $\forall i \quad 1 \le i \le N-1. \quad \text{Conn}(\text{Out}(O_\alpha, M_i), \text{In}(I_\alpha, N_{i,i+1}))$

(A4)  $\forall i \quad 1 \le i \le N-2, \quad \forall j \quad i+1 \le j \le N-1. \quad \text{Conn}(\text{Out}(O_\alpha, N_{i,j}), \text{In}(I_\alpha, N_{i,j+1}))$

(A5)  $\forall i \quad 1 \le i \le N-2, \quad \forall j \quad i+2 \le j \le N. \quad \text{Conn}(\text{Out}(O_\beta, N_{i,j}), \text{In}(I_\beta, N_{i+1,j}))$

(A6)  $\forall i \quad 1 \le i \le N-1. \quad \text{Conn}(\text{Out}(O_\beta, N_{i,i+1}), \text{In}(I_\beta, M_{i+1}))$

## 2. Behavior Specification:

### (a) Function specification:

(B1)  $\text{Conn}(X,Y) \wedge \text{Val}(X,x) \to \text{Val}(Y,x)$

(B2)  $\forall i \quad 1 \le i \le N. \quad I_{PM}(M_i) \wedge \text{Val}(\text{In}(I_\beta, M_i), \underline{c}_{input})$

$$\to \text{Val}(\text{Mem}(M_\beta, M_i), \bigwedge_{r=1}^{3i-2} \_ , \square \, \underline{c}_{input} \downarrow_{3i-2}) \wedge$$

$$\text{Val}(\text{Mem}(M_\alpha, M_i), \_ , \underline{c}_{input} / \bigwedge_{r=1}^{3i-2} \_ , \square \, \underline{c}_{input} \downarrow_{3i-2}) \wedge$$

$$\text{Val}(\text{Mem}(M_\gamma, M_i), \_ , \underline{c}_{input} - (\_ , \underline{c}_{input} / \bigwedge_{r=1}^{3i-2} \_ , \square \, \underline{c}_{input} \downarrow_{3i-2}) * (\bigwedge_{r=1}^{3i-2} \_ , \square \, \underline{c}_{input} \downarrow_{3i-2})) \wedge$$

$$\text{Val}(\text{Out}(O_\gamma, M_i), \_ , \underline{c}_{input} / \bigwedge_{r=1}^{3i-2} \_ , \square \, \underline{c}_{input} \downarrow_{3i-2}) \wedge$$

$$\text{Val}(\text{Out}(O_\alpha, M_i), \_ , \underline{c}_{input} / \bigwedge_{r=1}^{3i-2} \_ , \square \, \underline{c}_{input} \downarrow_{3i-2})$$

(B3) $\forall i \quad 1 \le i \le N-1, \quad \forall j \quad i+1 \le j \le N.$

$\qquad I_{PN}(N_{i,j}) \wedge \mathrm{Val}(\mathrm{In}(I_\alpha, N_{i,j}), \underline{a}_{input}) \wedge \mathrm{Val}(\mathrm{In}(I_\beta, N_{i,j}), \underline{c}_{input})$

$\qquad \to \ \mathrm{Val}(\mathrm{Mem}(M_\beta, N_{i,j}), \bigwedge_{r=1}^{2i+j-2} \_, \Box \underline{c}_{input} \downarrow_{2i+j-2}) \wedge$

$\qquad \mathrm{Val}(\mathrm{Mem}(M_\alpha, N_{i,j}), \_, \underline{a}_{input}) \wedge$

$\qquad \mathrm{Val}(\mathrm{Mem}(M_\gamma, N_{i,j}), (\_, \underline{c}_{input}) - (\_, \underline{a}_{input}) * (\bigwedge_{r=1}^{2i+j-2} \_, \Box \underline{c}_{input} \downarrow_{2i+j-2})) \wedge$

$\qquad \mathrm{Val}(\mathrm{Out}(O_\alpha, N_{i,j}), \_, \underline{a}_{input}) \wedge$

$\qquad \mathrm{Val}(\mathrm{Out}(O_\beta, N_{i,j}), (\_, \underline{c}_{input}) - (\_, \underline{a}_{input}) * (\bigwedge_{r=1}^{2i+j-2} \_, \Box \underline{c}_{input} \downarrow_{2i+j-2}))$

## (b) Input specifications:

(C1) $\mathrm{Val}(\mathrm{In}(I_\beta, M_1), \bigwedge_{r=1}^{N} c_{r,1}, \Box\_)$

(C2) $\forall j \quad 2 \le j \le N. \quad \mathrm{Val}(\mathrm{In}(I_\beta, N_{1,j}), \bigwedge_{r=1}^{j-1} \_, \bigwedge_{r=1}^{N} c_{r,j}, \Box\_)$

## (c) Output specifications:

(P1) $\forall i \quad 1 \le i \le N. \quad \mathrm{Val}(\mathrm{Out}(O_\gamma, M_i), \bigwedge_{r=1}^{3i-2} \_, \bigwedge_{r=i}^{N} FA(r,i,i), \Box\_)$

(P2) $\forall i \quad 1 \le i \le N. \quad \mathrm{Val}(\mathrm{Mem}(M_\beta, M_i), \bigwedge_{r=1}^{3i-2} \_, \Box FB(i,i,i))$

(P3) $\forall i \quad 1 \le i \le N-1, \quad \forall j \quad i+1 \le j \le N.$

$\qquad \mathrm{Val}(\mathrm{Mem}(M_\beta, N_{i,j}), \bigwedge_{r=1}^{2i+j-2} \_, \Box FB(i,j,i))$

LU Decomposition Equality Definitions:

$\qquad FC(i,j,0) = c_{i,j}$

$\qquad FC(i,j,k) = FC(i,j,k-1) - FA(i,j,k) * FB(i,j,k) \qquad \text{where } k > 0$

$\qquad FB(1,j,1) = c_{1,j}$

$\qquad FB(i,j,k) = FC(i,j,k-1) \qquad \text{where } i = k \text{ and } i > 1$

$\qquad FB(i,j,k) = FB(i-1,j,k) \qquad \text{where } i \ne k \text{ and } i > 1$

$\qquad FA(i,1,k) = FC(i,1,k-1) / FB(i,1,k)$

$\qquad FA(i,j,k) = FC(i,j,k-1) / FB(i,j,k) \qquad \text{where } j = k \text{ and } j > 1$

$$FA(i,j,k) = FA(i,j-1,k) \qquad \text{where} \quad j \neq k \text{ and } j > 1$$

Intermediate Specifications:

Specifications (T1), (T2), and (T3):

(T1) $\quad \forall i \quad 2 \leq i \leq N-1, \quad \forall j \quad i+1 \leq j \leq N.$

$$\text{Val}(\text{In}(I_\beta, N_{i,j}), \bigwedge_{r=1}^{2i+j-3} \_, \bigwedge_{r=i}^{N} FC(r,j,i-1), \square\_)$$

(T2) $\quad \forall i \quad 2 \leq i \leq N. \qquad \text{Val}(\text{In}(I_\beta, M_i), \bigwedge_{r=1}^{3i-3} \_, \bigwedge_{r=i}^{N} FC(r,i,i-1), \square\_)$

(T3) $\quad \forall i \quad 1 \leq i \leq N-1, \quad \forall j \quad i+1 \leq j \leq N.$

$$\text{Val}(\text{In}(I_\alpha, N_{i,j}), \bigwedge_{r=1}^{2i+j-3} \_, \bigwedge_{r=i}^{N} FA(r,i,i), \square\_)$$

## 7.3   PROOF OF CORRECTNESS OF THE ARRAY

To prove the correctness of the LU decomposition array, we need to prove three different sub-arrays $S_1$, $S_k$, and $S_n$, as shown in Figure 7.1. $S_1$ is considered as a different case due to different inputs, and $S_n$ is another different case since it does not consist of cells $N_{i,j}$ and the corresponding data flows. The 1-D array $S_1$ can be proved by mathematical induction given the inputs, the structural and the cell functional specifications. Based on the correctness of $S_1$ we can confirm the outputs from this $S_1$ sub-array, and hence the inputs to sub-array $S_k$. With the correct inputs and the given specifications, reverse induction is now used to prove sub-array $S_k$, because this array is trapezoidal, forming a part of the triangular array, as shown in Figure 7.1, with the longest row closest to the inputs. We can prove this longest row first and use it as the base for reverse induction. Finally, sub-array $S_n$ can be proved by trivial rewriting technique since only one cell is involved. This completes the proof. The following subsections give the details.

### 7.3.1  Proofs of the Intermediate Specifications

The proof begins with proving the three intermediate specifications, (T1), (T2), and (T3), as specified above. Due to the architecture and data flow of the array, (T1), (T2), and (T3) are difficult to prove. To prove (T1), one needs to prove (T3) first. To prove (T2), one needs to prove (T1) and (T3). And to prove (T3), one needs to prove (T1) and (T2). Since Specifications (T1), (T2), and (T3) mutually depend on each other, special heuristics need to be applied to the proofs of (T1), (T2), and (T3) to avoid vicious circle. Our solution is to transform Specification (T1) $\wedge$ (T2) $\wedge$ (T3) to their semantically identical form (S1) $\wedge$ (Sk) $\wedge$ (Sn), corresponding to sub-arrays $S_1$, $S_k$, and $S_n$, respectively. We first produce Specification (S1) from Specification (T3). Let $i = 1$:

Spec. (S1) $\quad \forall j \quad 2 \le j \le N. \quad \mathrm{Val}(\mathrm{In}(I_\alpha, N_{1,j}), \bigwedge_{r=1}^{j-1} \_, \bigwedge_{r=1}^{N} FA(r,1,1), \Box\_)$

Next, we produce Specification (Sk) from Specifications (T1), (T2), and (T3)

Spec. (Sk) $\quad \forall i \quad 2 \le i \le N-1, \quad \forall j \quad i+1 \le j \le N.$

$$\mathrm{Val}(\mathrm{In}(I_\beta, N_{i,j}), \bigwedge_{r=1}^{2i+j-3} \_, \bigwedge_{r=i}^{N} FC(r,j,i-1), \Box\_) \wedge$$

$$\mathrm{Val}(\mathrm{In}(I_\beta, M_i), \bigwedge_{r=1}^{3i-3} \_, \bigwedge_{r=i}^{N} FC(r,i,i-1), \Box\_) \wedge$$

$$\mathrm{Val}(\mathrm{In}(I_\alpha, N_{i,j}), \bigwedge_{r=1}^{2i+j-3} \_, \bigwedge_{r=i}^{N} FA(r,i,i), \Box\_)$$

We now generate Specification (Sn) from Specification (T2). Let $i = N$:

Spec. (Sn) $\quad \mathrm{Val}(\mathrm{In}(I_\beta, M_N), \bigwedge_{r=1}^{3N-3} \_, FC(r,N,N-1), \Box\_)$

Specification (S1) is proved by induction on the number of N cells. Based on the proof, corollary (S1) is also derived:

Corollary (S1)

$$\forall j \quad 2 \le j \le N. \quad \mathrm{Val}(\mathrm{Out}(O_\beta, N_{1,j}), \bigwedge_{r=1}^{j+1} \_, \bigwedge_{r=2}^{N} FC(r,j,1), \Box\_)$$

Specification (Sk) is proved by reverse induction on the length of rows of the array. The base case and the induction step both invoke another level of simple induction. Corollary (Sk) is derived from Specification (Sk), and is used to prove Specification (Sn), which is done by unfolding and rewriting.

Corollary (Sk)

$$\mathrm{Val}(\mathrm{Out}(O_\beta, N_{N-1,N}), \bigwedge_{r=1}^{3N-3} \_, FC(r,N,N-1), \Box\_)$$

The proof of Specification (Sk) is briefly described below:

**Base case:**

$$i = 2, \quad \forall j \quad 3 \le j \le N.$$

$$\text{Val}(\text{In}(I_\beta, N_{2,j}), \bigwedge_{r=1}^{j+1} \_ , \bigwedge_{r=2}^{N} FC(r,j,1), \square\_) \wedge$$

$$\text{Val}(\text{In}(I_\beta, M_2), \bigwedge_{r=1}^{3} \_ , \bigwedge_{r=2}^{N} FC(r,2,1), \square\_) \wedge$$

$$\text{Val}(\text{In}(I_\alpha, N_{2,j}), \bigwedge_{r=1}^{j+1} \_ , \bigwedge_{r=2}^{N} FA(r,2,2), \square\_)$$

Let these three value predicates be (G1), (G2), and (G3), respectively. From Specification (A6),

$$i = 1. \quad \text{Conn}(\text{Out}(O_\beta, N_{1,2}), \text{In}(I_\beta, M_2))$$

(G2) becomes

$$\text{Val}(\text{Out}(O_\beta, N_{1,2}), \bigwedge_{r=1}^{3} \_ , \bigwedge_{r=2}^{N} FC(r,2,1), \square\_)$$

By (A5),

$$i = 1, \quad \forall j \quad 3 \le j \le N. \quad \text{Conn}(\text{Out}(O_\beta, N_{1,j}), \text{In}(I_\beta, N_{2,j}))$$

(G1) becomes

$$\text{Val}(\text{Out}(O_\beta, N_{1,j}), \bigwedge_{r=1}^{j+1} \_ , \bigwedge_{r=2}^{N} FC(r,j,1), \square\_)$$

Combining new (G1) and new (G2), we have

$$\forall j \quad 2 \le j \le N. \quad \text{Val}(\text{Out}(O_\beta, N_{1,j}), \bigwedge_{r=1}^{j+1} \_ , \bigwedge_{r=2}^{N} FC(r,j,1), \square\_)$$

Which is the same as Corollary (S1), hence the base case proof of (G1) and (G2) is complete.

To prove (G3), we perform a mathematical induction on the length of the row. Finally, the base case proof of Specification (Sk) is done.

**Induction step:**

Hypothesis:

$i = k, \quad \forall j \quad k+1 \leq j \leq N.$

$$\text{Val}(\text{In}(I_\beta, N_{k,j}), \bigwedge_{r=1}^{2k+j-3} \_, \bigwedge_{r=k}^{N} FC(r, j, k-1), \square \_) \wedge$$

$$\text{Val}(\text{In}(I_\beta, M_k), \bigwedge_{r=1}^{3k-3} \_, \bigwedge_{r=k}^{N} FC(r, k, k-1), \square \_) \wedge$$

$$\text{Val}(\text{In}(I_\alpha, N_{k,j}), \bigwedge_{r=1}^{2k+j-3} \_, \bigwedge_{r=k}^{N} FA(r, k, k), \square \_)$$

Let these three value predicates be (H1), (H2), and (H3) in their order respectively,

Goal:

$i = k+1, \quad \forall j \quad k+2 \leq j \leq N.$

$$\text{Val}(\text{In}(I_\beta, N_{k+1,j}), \bigwedge_{r=1}^{2k+j-1} \_, \bigwedge_{r=k+1}^{N} FC(r, j, k), \square \_) \wedge$$

$$\text{Val}(\text{In}(I_\beta, M_{k+1}), \bigwedge_{r=1}^{3k} \_, \bigwedge_{r=k+1}^{N} FC(r, k+1, k), \square \_) \wedge$$

$$\text{Val}(\text{In}(I_\alpha, N_{k+1,j}), \bigwedge_{r=1}^{2k+j-1} \_, \bigwedge_{r=k+1}^{N} FA(r, k+1, k+1), \square \_)$$

Let these three value predicates be (G1'), (G2'), and (G3'), respectively. Using (A5), (A6), and combining (G1') and (G2'), (G1') and (G2') are proved similarly to the base case of (G1) and (G2) above, except that (A2), (H1), and (H3) are used. Mathematical induction is used again to prove goal (G3'). Thus, the induction step proof of Specification (Sk) is done.

Finally, Specification (Sn) is proved by using (A6) and Corollary (Sk). The proof of (S1) ∧ (Sk) ∧ (Sn) is thus complete.


## 7.3.2 Proofs of the Output Specifications

This section presents the proofs of Output Specifications (P1), (P2), and (P3), which constitute the final step in our formal verification process. Proof of (P1) is done by case analysis.

$$(\text{P1}) \quad \forall i \quad 1 \leq i \leq N. \quad \text{Val}(\text{Out}(O_\gamma, M_i), \bigwedge_{r=1}^{3i-2} \_, \bigwedge_{r=i}^{N} FA(r, i, i), \square \_)$$

*Case 1:*

$$i = 1. \quad \text{Val}(\text{Out}(O_\gamma, M_1), \bigwedge_{r=1}^{1} \_, \bigwedge_{r=1}^{N} FA(r, 1, 1), \square \_)$$

By (B2), let $i = 1$, and by (A1) and (C1), we have

$$\underline{c}_{input} = \bigwedge_{r=1}^{N} c_{r,1}, \square_{\_}$$

Hence we have

$$\mathrm{Val}(\mathrm{Out}(O_\gamma, M_1), (\_, \bigwedge_{r=1}^{N} c_{r,1}, \square_{\_}) / (\bigwedge_{r=1}^{1} \_, \square\, c_{1,1}))$$

Hence

$$\mathrm{Val}(\mathrm{Out}(O_\gamma, M_1), \_, \bigwedge_{r=1}^{N} c_{r,1} / c_{1,1}, \square_{\_})$$

Hence

$$\mathrm{Val}(\mathrm{Out}(O_\gamma, M_1), \_, \bigwedge_{r=1}^{N} FA(r,1,1), \square_{\_})$$

Hence, Case 1 proof is complete.

*Case 2*:

$$\forall i \quad 2 \leq i \leq N. \quad \mathrm{Val}(\mathrm{Out}(O_\gamma, M_i), \bigwedge_{r=1}^{3i-2} \_, \bigwedge_{r=i}^{N} FA(r,i,i), \square_{\_})$$

By (B2), (A1), and (T2), we have

$$\underline{c}_{input} = \bigwedge_{r=1}^{3i-3} \_, \bigwedge_{r=i}^{N} FC(r,i,i-1), \square_{\_}$$

Hence we have

$$\forall i \quad 2 \leq i \leq N. \quad \mathrm{Val}(\mathrm{Out}(O_\gamma, M_i), (\_, \bigwedge_{r=1}^{3i-3} \_, \bigwedge_{r=i}^{N} FC(r,i,i-1), \square_{\_}) / (\bigwedge_{r=1}^{3i-2} \_, \square FC(i,i,i-1)))$$

Hence

$$\forall i \quad 2 \leq i \leq N. \quad \mathrm{Val}(\mathrm{Out}(O_\gamma, M_i), \bigwedge_{r=1}^{3i-2} \_, \bigwedge_{r=i}^{N} FC(r,i,i-1) / FC(i,i,i-1), \square_{\_})$$

Hence (from equality definitions)

$$\forall i \quad 2 \leq i \leq N. \quad \text{Val}(\text{Out}(O_\gamma, M_i), \bigwedge_{r=1}^{3i-2} \_ , \bigwedge_{r=i}^{N} FA(r,i,i), \square\_)$$

Hence Case 2 proof is complete.

Proof (P2) and Proof (P3) are also proved by case analysis, which are similar to Proof (P1). The abstract proof trees of the LU decomposition example are given in Figure 7.2. We give only the abstract proof trees here. Each node in the abstract proof tree explains a proof sequence and the facts or rules used in the proof.

## 7.4   CONCLUSION

The executions of the proofs for different systolic designs were performed in SICStus Prolog on our Sun workstations. A sample session of an automated formal proof of correctness of a 2-D matrix-multiplication systolic design is listed in Appendix A. It takes a total elapsed time of less than 5 minutes, and an execution time of less than 30 seconds, if excluding time of interaction.

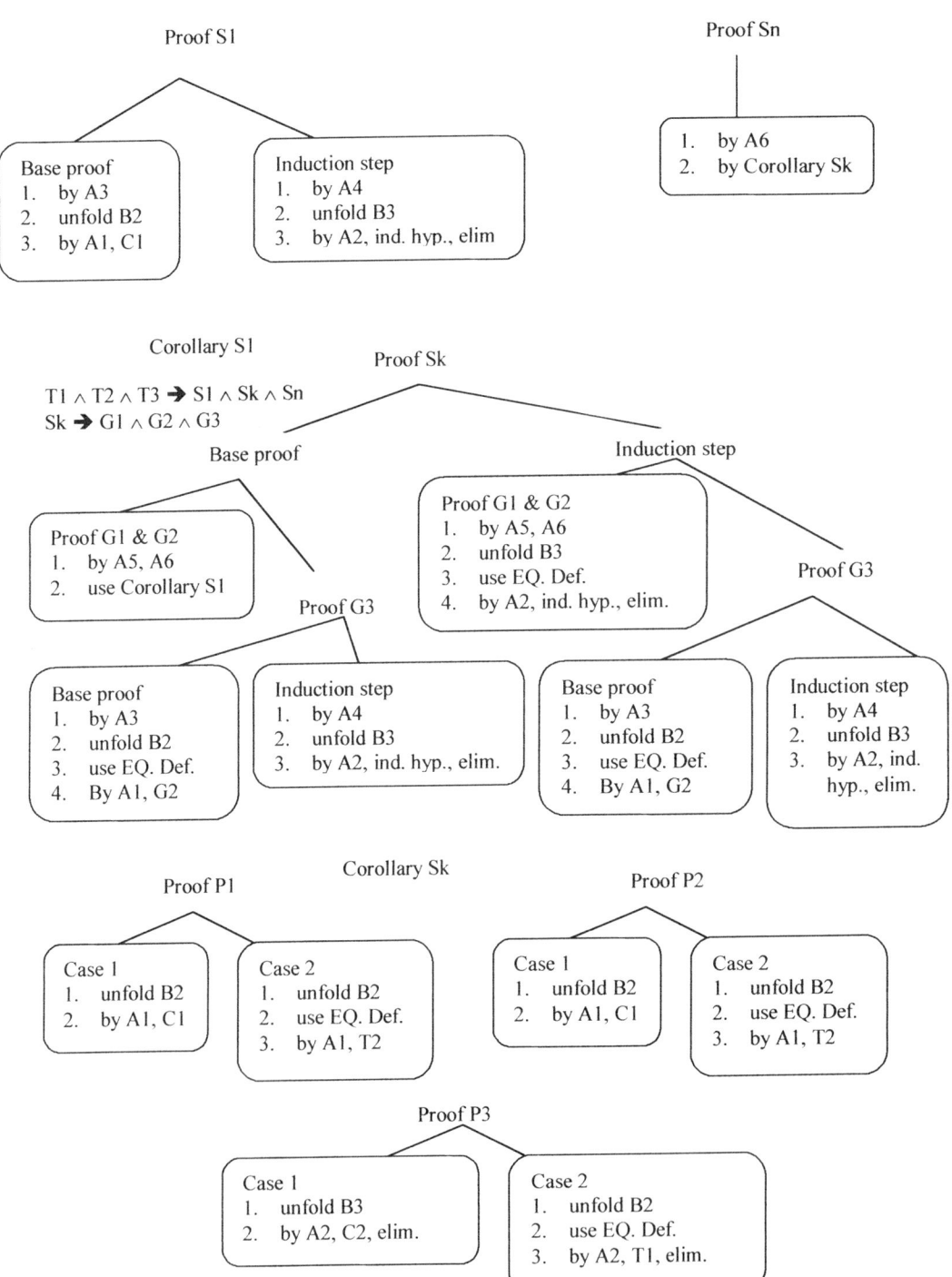

Figure 7.2 Abstract sub-proof trees showing the verification process for the triangular
LU decomposition systolic array

# Chapter 8

# CONCLUSIONS

Systolic arrays offer both parallelism and pipelining required for high-speed computation in many signal processing, image processing, as well as linear algebraic applications. Although many methods have been developed to produce correct-by-construction arrays from given algorithms, these methods can only be applied to a limited class of algorithms. In fact, many systolic arrays are designed by ad hoc or systematic, but not necessarily formal, techniques. Formal methods are therefore necessary to serve as tools to guarantee the correctness of systolic designs, since simulation may leave many design errors undetected. It is in this direction that we developed the tools described in this book. With efficient formal specification and verification tools for systolic architecture, precise and complete description of designs as well as ensuring their correctness can be made possible and design and debugging time can be significantly reduced.

We briefly summarize our research contribution in this chapter. Firstly, a novel formalism, named *Systolic Temporal Arithmetic* (*STA*), is developed. Useful STA axioms, rules, and theorems are also presented and proved. The formalism exploits systolic properties to provide constructs and verification techniques for effective and efficient specification and verification of systolic array designs. Secondly, a framework for formal specification and verification of systolic arrays is produced. Three verification strategies, output derivation and comparison, induction, and solving STA difference equations, are generated. Several application examples are provided in the book to illustrate our framework and verification techniques. Finally, a Prolog-based verifier, named *Verifier for STA* (*VSTA*), is built to provide a CAD facility for semi-automated interactive design verification. We also provide an example showing how a complex systolic array designed for LU decomposition can be specified and verified by our technique.

As major industries are presently incorporating formal verification methods into their CAD systems, our contributions offer solutions for regular architectures like systolic arrays. Possible future research includes the integration of our results with existing circuit level verification techniques and incorporating them into design packages.

## A.1 VSTA USER INTERFACE

The interface of VSTA is running under the Motif/X11 windowing environment [Shih95]. Users may enter their design specifications in separate text windows (structural specification window, functional specification window, etc.). Figure A.1 shows specification editor windows. To start a verification process, one brings up a verification window. Figure A.2 shows a verification window with a sample proof trace. VSTA may ask for necessary decisions during the verification process.

## A.2 VSTA SAMPLE SPECIFICATION

A sample VSTA design specification for our matrix-matrix multiplication 2-D systolic array is given here. The array is described in Section 5.2 and shown again in Figure A.3. STA specification is provided in Prolog. Operator definitions are provided first, followed by STA specification.

```
%%% STA temporal variable operator definitions

%%% temporal variable external representation
%%%   tv:List:L is a temporal variable of length L. List is a list
%%%   of the first L normal variables in the temporal variable
:- op(600, xfy, :).

%%% next operator:
%%%   N o T is a temporal variable obtained by normal variable N
and
%%%   temporal variable T
:- op(610, xfy, o).

%%% temporal variable substitution:
%%%   X $ Y is a temporal variable obtained by substituting the i
th
%%%   normal variable of Y for the i th normal variable of X,
where
%%%   i is the position of the first unknow normal variable in X.
```

```
%%%    That is, the difference between temporal variable X $ Y and
%%%    temporal variable X is only at the i th normal variable.
:- op(615, xfy, $).

%%% temporal variable concatenation:
%%%    X ^ Y is a temporal variable obtained by concatenating
temporal
%%%    variables X and Y
:- op(620, yfx, ^).

%%% temporal variable multiplication:
%%% X ** Y is the temporal product of temporal variables X and Y
:- op(630, yfx, **).

%%% temporal variable division:
%%% X // Y is the temporal quotient of temporal variables X and Y
:- op(630, yfx, //).

%%% temporal variable addition:
%%% X ++ Y is the temporal sum of temporal variables X and Y
:- op(640, yfx, ++).

%%% temporal variable subtraction:
%%% X -- Y is the temporal difference of temporal variables X and
Y
:- op(640, yfx, --).

%%% temporal variable assignment:
%%% V := T assigns the temporal variable T a name V
:- op(650, xfx, :=).

%%% bound universal quantifier:
%%% forall 1 <= I <= N suchthat A
:- op(610,  fx, forall).
:- op(600, yfx, <=).
:- op(600, yfx, <).

%%% implication:
%%% Hyps implies Concl
:- op(630, xfx, implies).

%%% temporal variable projection:
%%% TV \ i
:- op(615, yfx, \).

%%% equality:
%%% WFF equals WFF
```

```
:- op(630, xfx, equals).

%%% such that
%%% WFFa suchthat WFFb
:- op(620, xfx, suchthat).

%%% concatenation
%%% conc Idx-LB-UB-Parm-Term
:- op(550, fx, conc).
:- op(550, fx, conc1).

%%% summation
%%% sum Idx-LB-UB-Parm-Term
:- op(550, fx, sum).

%%% accumulation
%%% acc Idx-LB-UB-Parm-Term
:- op(550, fx, acc).


%%% STA Specification Example of Matrix-Matrix Multiplication
Systolic Array

%%% Structure Specifications

%%% Specification A1
(sa_size(N), forall 1 <= I <= N, forall 1 <= J <= N)
  suchthat
  ip(y:(I, J)).

%%% Specification A2
(sa_size(N), forall 1 <= I <= N, forall 1 <= J < N)
  suchthat
  conn(out(o:alpha, y:(I, J)), in(i:alpha, y:(I, J+1))).

%%% Specification A3
(sa_size(N), forall 1 <= J <= N, forall 1 <= I < N)
  suchthat
  conn(out(o:beta, y:(I, J)), in(i:beta, y:(I+1, J))).


%%% Behavior Specifications

%%% Specification B1
( ( sa_size(N),
    nof_row(NofR),
    forall 1 <= I <= N,
```

```
          forall 1 <= J <= N) suchthat
     ( ip(y:(I, J)),
          val(in(i:alpha, y:(I, J)), Ain),
          val(in(i:beta, y:(I, J)),  Bin),
          val(mem(m:alpha, y:(I, J)),  Y) ) )
     implies
          val(mem(m:alpha, y:(I, J)), $ o ( Y ++ sum L-1-NofR-(I+J+L)-
               (conc R-1-(I+J+L-3)-0 ^ tv:[Ain \ L * Bin \ L]:1)) ).


%%% Specification B13
( ( sa_size(N),
     nof_row(NofR),
     forall 1 <= I < N,
     forall 1 <= J < N) suchthat
     ( ip(y:(I, J)),
          val(in(i:alpha, y:(I, J)),
               tv:[0]:I-2+J  ^
               conc R-1-NofR-I-(a:[I, R])   ^
               tv:[0]:1 ),
          val(in(i:beta, y:(I, J)),
               tv:[0]:J-2+I   ^
               conc R-1-NofR-J-(b:[R, J])   ^
               tv:[0]:1 ),
          val(mem(m:alpha, y:(I, J)),
               tv:[0]:I+J-1   ^
               conc R-1-NofR-(I+J+R)-(sum M-1-R-(I+J)-
                    (a:[I, M] * b:[M, J]))   ^
               tv:[0]:1 ) ) )
     implies
          val(mem(m:alpha, y:(I+1, J+1)),
               tv:[0]:(I+1)+(J+1)-1   ^
               conc R-1-NofR-((I+1)+(J+1)+R)-(sum M-1-R-
                    ((I+1)+(J+1))-(a:[I+1, M] * b:[M, J+1])) ^
               tv:[0]:1 ).


%%% Specification B11
( ( sa_size(N),
     nof_row(NofR),
     forall 1 <= I <  N,
     forall 1 <= J <= N) suchthat
     ( ip(y:(I, J)),
          val(in(i:alpha, y:(I, J)),
               tv:[0]:I-2+J  ^
               conc R-1-NofR-I-(a:[I, R])   ^
               tv:[0]:1 ),
          val(in(i:beta, y:(I, J)),
               tv:[0]:J-2+I   ^
```

```
           conc R-1-NofR-J-(b:[R, J])   ^
           tv:[0]:1 ),
     val(mem(m:alpha, y:(I, J)),
           tv:[0]:I+J-1   ^
           conc R-1-NofR-(I+J+R)-(sum M-1-R-(I+J)-
               (a:[I, M] * b:[M, J]))   ^
           tv:[0]:1 ) ) )
   implies
     val(mem(m:alpha, y:(I+1, J)),
           tv:[0]:(I+1)+J-1   ^
           conc R-1-NofR-((I+1)+J+R)-(sum M-1-R-((I+1)+J)-
               (a:[I+1, M] * b:[M, J])) ^
           tv:[0]:1 ).

%%% Specification B12
( ( sa_size(N),
     nof_row(NofR),
     forall 1 <= I <= N,
     forall 1 <= J <  N) suchthat
   ( ip(y:(I, J)),
     val(in(i:alpha, y:(I, J)),
           tv:[0]:I-2+J   ^
           conc R-1-NofR-I-(a:[I, R])   ^
           tv:[0]:1 ),
     val(in(i:beta, y:(I, J)),
           tv:[0]:J-2+I   ^
           conc R-1-NofR-J-(b:[R, J])   ^
           tv:[0]:1 ),
     val(mem(m:alpha, y:(I, J)),
           tv:[0]:I+J-1   ^
           conc R-1-NofR-(I+J+R)-(sum M-1-R-(I+J)-
               (a:[I, M] * b:[M, J]))   ^
           tv:[0]:1 ) ) )
   implies
     val(mem(m:alpha, y:(I, J+1)),
           tv:[0]:I+(J+1)-1   ^
           conc R-1-NofR-(I+(J+1)+R)-(sum M-1-R-(I+(J+1))-
               (a:[I, M] * b:[M, J+1])) ^
           tv:[0]:1 ).

%%% Specification D1
(sa_size(N), forall 1 <= I <= N, forall 1 <= J <= N)
   suchthat
   val(in(i:alpha, y:(I, J)), Ain) implies
   val(out(o:alpha, y:(I, J)), 0 o Ain).

%%% Specification D2
```

```
(sa_size(N), forall 1 <= I <= N, forall 1 <= J <= N)
  suchthat
  val(in(i:beta, y:(I, J)), Bin) implies
  val(out(o:beta, y:(I, J)), 0 o Bin).

%%% Specification B2
( conn(X, Y),
  val(X, Xval) ) implies
  val(Y, Xval).

%%% Specification C1
(sa_size(N), nof_row(NofR), forall 1 <= I <= N)
  suchthat
  val(in(i:alpha, y:(I, 1)),
      tv:[0]:I-2+1   ^
      conc R-1-NofR-I-(a:[I, R])   ^
      tv:[0]:1).

%%% Specification C2
(sa_size(N), nof_row(NofR), forall 1 <= J <= N)
  suchthat
  val(in(i:beta, y:(1, J)),
      tv:[0]:J-2+1   ^
      conc R-1-NofR-J-(b:[R, J])   ^
      tv:[0]:1).

%%% Specification C3
(sa_size(N), forall 1 <= I <= N, forall 1 <= J <= N)
  suchthat
  val(mem(m:alpha, y:(I, J)), tv:[0]:1).

%%% Specification P11
(sa_size(N), nof_row(NofR), forall 1 <= I <= N)
  suchthat
  val(in(i:alpha, y:(I, N)),
      tv:[0]:I-2+N   ^
      conc R-1-NofR-I-(a:[I, R])   ^
      tv:[0]:1).

%%% Specification P12
(sa_size(N), nof_row(NofR), forall 1 <= J <= N)
  suchthat
  val(in(i:beta, y:(N, J)),
      tv:[0]:J-2+N   ^
      conc R-1-NofR-J-(b:[R, J])   ^
      tv:[0]:1).
```

```
%%% Specification P1
(sa_size(N), nof_row(NofR),
  forall 1 <= I <= N, forall 1 <= J <= N) suchthat
    val(mem(m:alpha, y:(I, J)),
  tv:[0]:I+J-1   ^
  conc R-1-(NofR)-(I+J+R)-(sum M-1-R-(I+J)-
    (a:[I, M] * b:[M, J]))   ^
  tv:[0]:1).
```

## A.3 SAMPLE VSTA PROOF OUTPUT

A sample VSTA proof output for our matrix-matrix multiplication 2-D systolic array is given in this section. Its description and STA specification are provided in Section A.2.

The proof of this 2-D example consists of two inductive proofs (for specifications P11 and P12) and one double inductive proof (for specification P1). The first part of the proof shows an inductive proof of P11. Proof of P12 is done in a similar manner. Proof trees are given in Figure A.4 (a) – (c). In the proof, two sub-proofs, P11 and P12, are proved before the main proof P1 is proceeded. Sub-proof P11 shows the intermediate values at the input ports of cell y's in the horizontal positions. Sub-proof P11 is performed by regular induction, which consists of a base proof and an induction step proof. Each proof uses input specifications, structural specifications, and other rules in the nodes of the proof tree. Sub-proof P12, proved similarly by regular induction, shows the intermediate values of ports in the vertical positions. The main proof P1 is pursued by double induction in both the horizontal direction as well as in the vertical direction. Basically, the inductive step is to prove the correctness of the array of size ($k$+1) x ($k$+1), assuming a correct array of size $k$ x $k$. The base case is proved by an unfolding using specification B1. The induction step consists of three cases (fixing column indices, fixing row indices, and extending both indices). Each case is performed by unfolding using different specifications.

The complete proof output from VSTA is quite lengthy. A highly simplified sample is presented here (SICStus Prolog is used).

```
| ?- [sta2].
{consulting /users1/tshih/sta/sta2.pl...}
......
yes
| ?- prove.
***************************
***   Starting Sub Proof1   ***
***************************
%%%%%%%%%%%%%%%%%%%
%%% Base Case %%%
%%%%%%%%%%%%%%%%%%%
%%% Try Help predicate:
sa_size(1),nof_row(_177),forall 1<=_185<=1
```

```
     SUCH THAT
val(in(i:alpha,y:(_185,1)),tv:[0]:_185-2+1^conc _233-1-
_177-_185-(a:[_185,_233])^tv:[0]:1)


*** Help predicate succeeds
%%%%%%%%%%%%%%%%%%%%%%%
%%% Base Case Done %%%
%%%%%%%%%%%%%%%%%%%%%%%
%%%%%%%%%%%%%%%%%%%%%%%
%%% Induction Step %%%
%%%%%%%%%%%%%%%%%%%%%%%
%%% Try Constrained goal:
sa_size(k+1),nof_row(_674),forall 1<=_682<=k+1
  SUCH THAT
val(in(i:alpha,y:(_682,k+1)),tv:[0]:_682-2+(k+1)^conc
_730-1-_674-_682-(a:[_682,_730])^tv:[0]:1)

%%% Try Conjunction goal:
sa_size(k+1)

nof_row(_674)

forall 1<=_682<=k+1

%%% Try Help predicate:
sa_size(k+1)

*** Help predicate succeeds
%%% Try Conjunction goal:
nof_row(_674)

forall 1<=_682<=k+1

%%% Try Help predicate:
nof_row(4)

*** Help predicate succeeds
%%% Try Help predicate:
forall 1<=_682<=k+1

*** Help predicate succeeds
*** Conjunction goal succeeds
*** Conjunction goal succeeds
%%% Try Value predicate:
val(in(i:alpha,y:(_682,k+1)),tv:[0]:_682-2+(k+1)^conc
_730-1-4-_682-(a:[_682,_730])^tv:[0]:1)
```

```
*** Value predicate fails
%%% Try Valid fact:
val(in(i:alpha,y:(_682,k+1)),tv:[0]:_682-2+(k+1)^conc
_730-1-4-_682-(a:[_682,_730])^tv:[0]:1)
......
*** Valid fact fails
%%% Try Unfolding:
val(in(i:alpha,y:(_682,k+1)),tv:[0]:_682-2+(k+1)^conc
_730-1-4-_682-(a:[_682,_730])^tv:[0]:1)

*** Rule:
(sa_size(_2733),forall 1<=_2744<=_2733,forall 1<=_2752
<=_2733) suchthat (ip(y:(_2744,_2752)),val(in(i:alpha,
y:(_2744,_2752)),_2769),val(in(i:beta,y:(_2744,_2752)),
_2787),val(mem(m:alpha,y:(_2744,_2752)),_2802))
  IMPLIES
val(mem(m:alpha,y:(_2744,_2752)),f(_2769,_2787,_2802))
Fire this rule? (yes./no.) no.
......
*** Rule:
conn(_2730,_2731),val(_2730,_2734)
  IMPLIES
val(_2731,_2734)
Fire this rule? (yes./no.) yes.

%%% Try Conjunction goal:
conn(_2730,in(i:alpha,y:(_682,k+1)))

val(_2730,tv:[0]:_682-2+(k+1)^conc _730-1-4-_682-(a:[
_682,_730])^tv:[0]:1)
......
*** Unfolding succeeds
*** Constrained goal succeeds
%%%%%%%%%%%%%%%%%%%%%%%%%%%%%
%%% Induction Step Done %%%
%%%%%%%%%%%%%%%%%%%%%%%%%%%%%
%%%%%%%%%%%%%%%
%%% Q.E.D. %%%
%%%%%%%%%%%%%%%
*************************
***  Starting Sub Proof2  ***
*************************

%%%%%%%%%%%%%%%%%%%
%%% Base Case %%%
%%%%%%%%%%%%%%%%%%%
..........
Similarly, P12 is done by mathematical induction
```

```
. . . . . . . . . .
%%%%%%%%%%%%%%%%%%%%%%%%%%%%%
%%% Induction Step Done %%%
%%%%%%%%%%%%%%%%%%%%%%%%%%%%%
%%%%%%%%%%%%%%%%
%%% Q.E.D. %%%
%%%%%%%%%%%%%%
***************************
***   Starting Main Proof   ***
***************************
%%%%%%%%%%%%%%%%%%%
%%% Base Case %%%
%%%%%%%%%%%%%%%%%%%
%%% Try Constrained goal:
sa_size(_11578),nof_row(_11583),forall 1<=1<=_11578,
forall 1<=1<=_11578
   SUCH THAT
val(mem(m:alpha,y:(1,1)),tv:[0]:1+1-1^conc _11650-1-
_11583-(1+1+_1165 0)-(sum _11670-1-_11650-(1+1)-
(a:[1,_11670]*b:[_11670,1]))^tv:[0]:1)

%%% Try Conjunction goal:
sa_size(_11578)
. . . . . .
*** Constrained goal succeeds
*** Unfolding succeeds
*** Constrained goal succeeds
%%%%%%%%%%%%%%%%%%%%%%%%
%%% Base Case Done %%%
%%%%%%%%%%%%%%%%%%%%%%%%
%%%%%%%%%%%%%%%%%%%%%%%%
%%% Induction Step %%%
%%%%%%%%%%%%%%%%%%%%%%%%
%%% Try Constrained goal:
sa_size(_19683),nof_row(_19688),forall 1<=k+1<=_19683,
forall 1<=k<=_19683
   SUCH THAT
val(mem(m:alpha,y:(k+1,k)),tv:[0]:k+1+k-1^conc _19755-
1-_19688-(k+1+k+_19755)-(sum _19775-1-_19755-(k+1+k)-
(a:[k+1,_19775]*b:[_19775,k]))^tv:[0]:1)
. . . . . .
*** Constrained goal succeeds
%%%%%%%%%%%%%%%%%%%%%%%%%%%%%
%%% Induction Step Done %%%
%%%%%%%%%%%%%%%%%%%%%%%%%%%%%
%%%%%%%%%%%%%%%%%%%%%%%%
%%% Induction Step %%%
```

```
%%%%%%%%%%%%%%%%%%%%%%%
..........
The proof fixing another index is done in the same
manner.
..........
%%%%%%%%%%%%%%%%%%%%%%%%%%%%
%%% Induction Step Done %%%
%%%%%%%%%%%%%%%%%%%%%%%%%%%%
%%%%%%%%%%%%%%%%%%%%%%%
%%% Induction Step %%%
%%%%%%%%%%%%%%%%%%%%%%%
..........
The proof inducts on both index is done in the same
manner
..........
%%%%%%%%%%%%%%%%%%%%%%%%%%%%%
%%% Induction Step Done %%%
%%%%%%%%%%%%%%%%%%%%%%%%%%%%%
%%%%%%%%%%%%%%%
%%% Q.E.D. %%%
%%%%%%%%%%%%%%%
yes
| ?- halt.
```

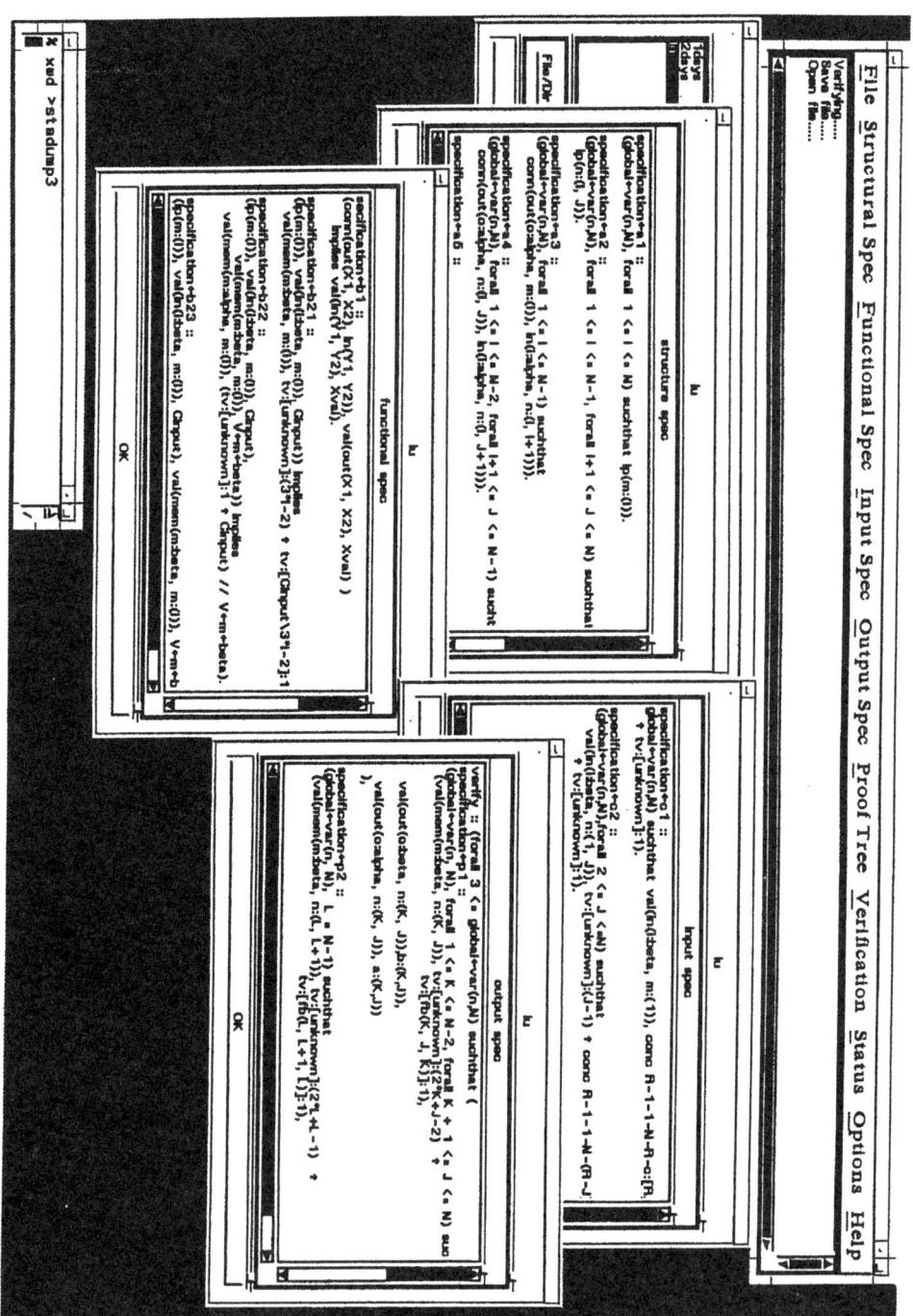

Figure A.1 VSTA specification editor windows

# VSTA Verification Window

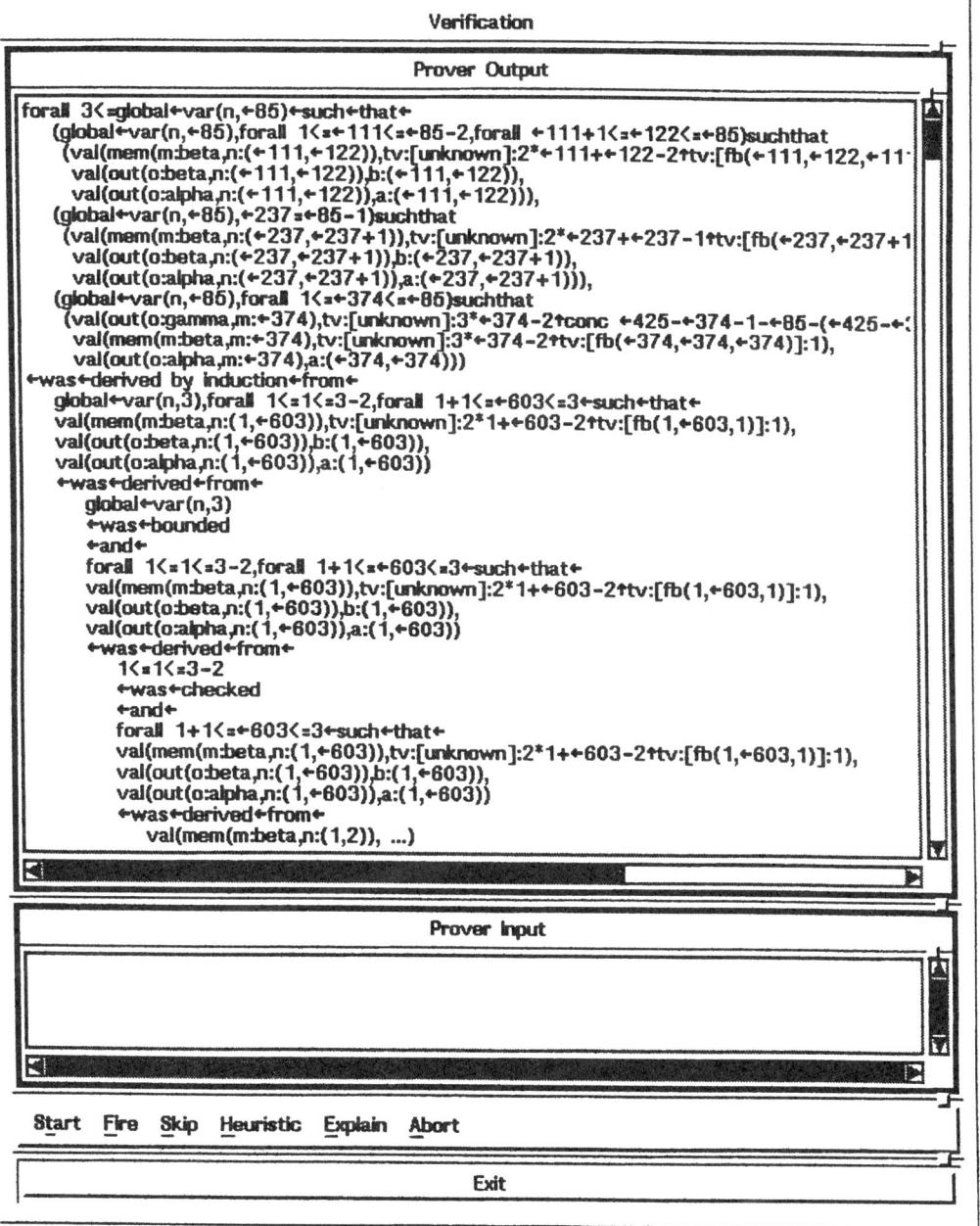

Figure A.2 VSTA verification window

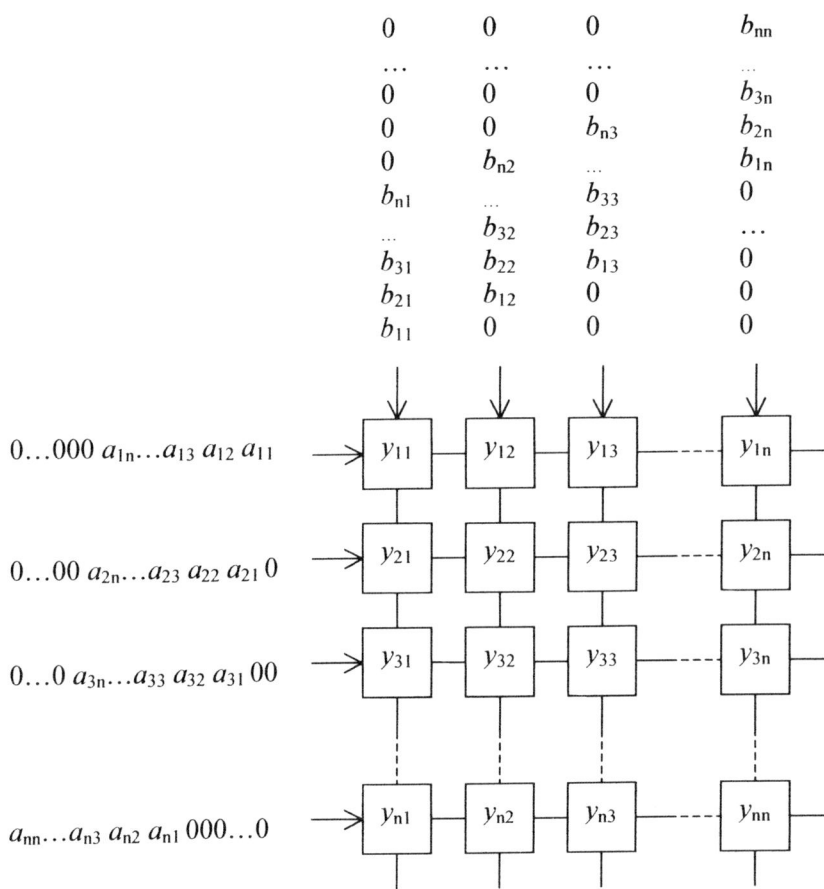

Cell definition:

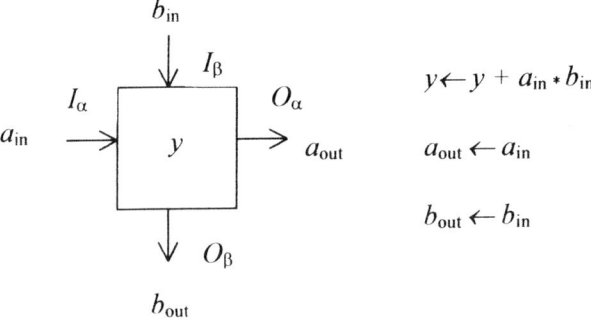

$$y \leftarrow y + a_{in} * b_{in}$$

$$a_{out} \leftarrow a_{in}$$

$$b_{out} \leftarrow b_{in}$$

Figure A.3 2-D systolic array for matrix-matrix multiplication and its cell definition

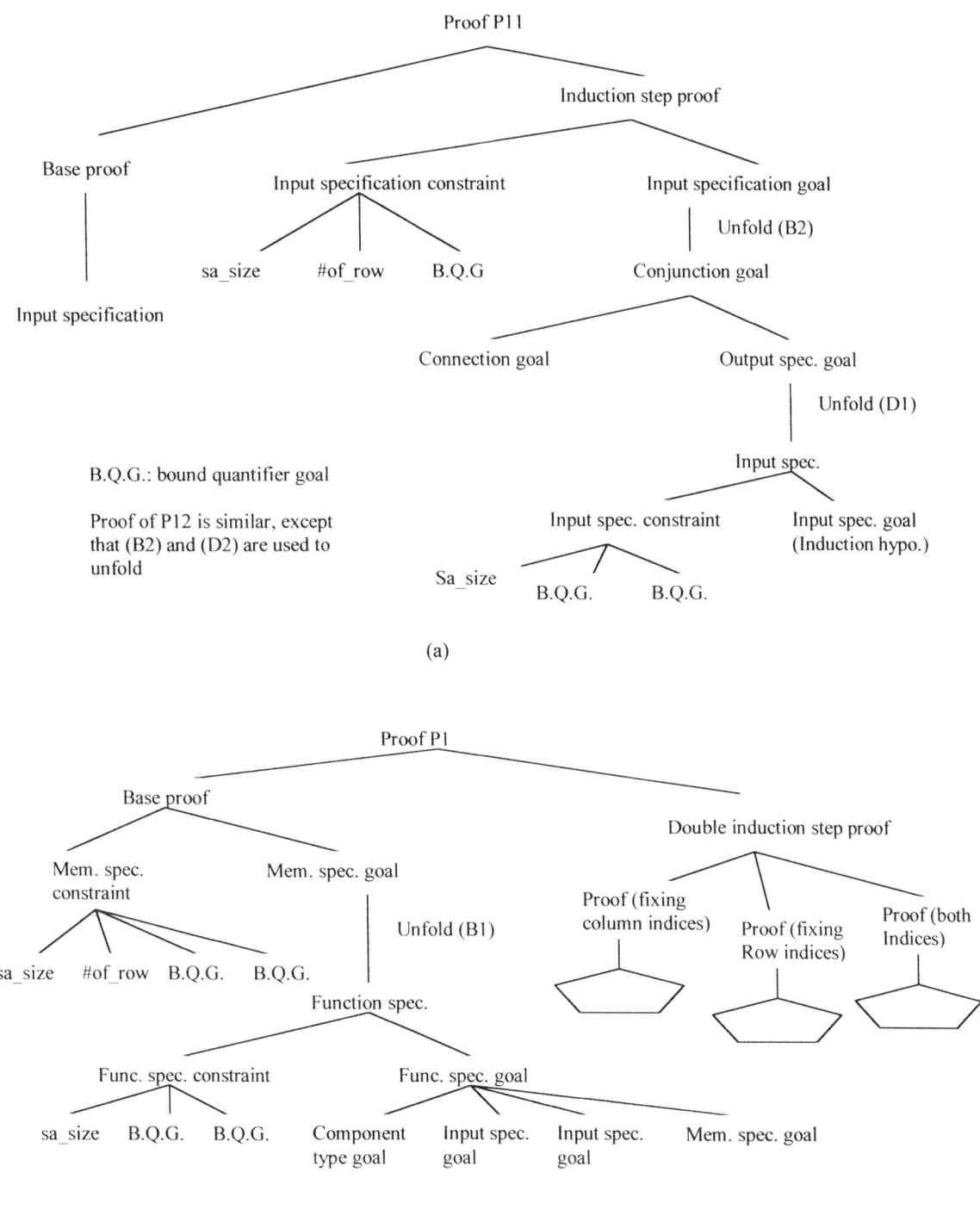

Figure A.4 (a)-(b) Proof trees for the matrix-matrix multiplication array

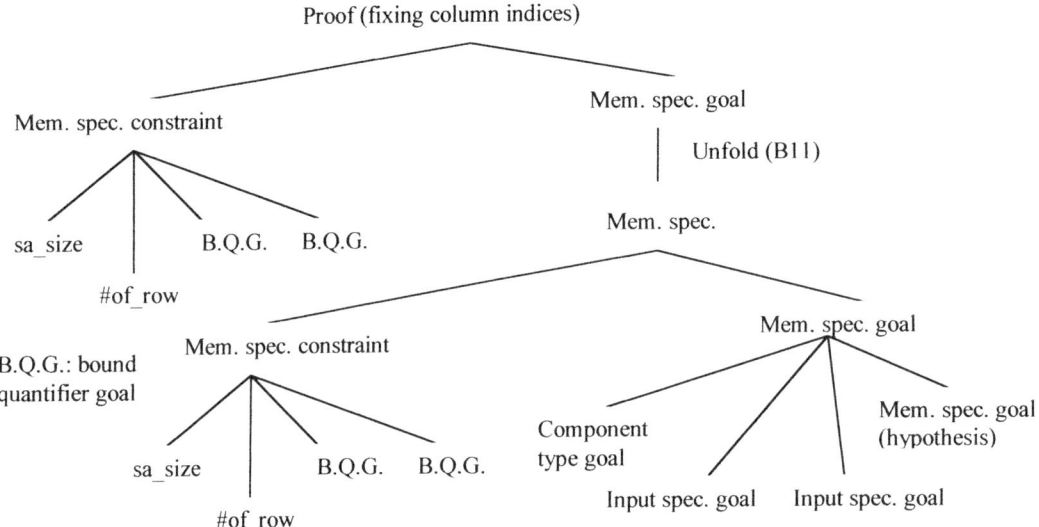

Figure A.4 (c) A sub-proof tree for the matrix-matrix multiplication array

# BIBLIOGRAPHY

[Abad86]    M.Abadi and Z.Manna, "A Timely Resolution", *Report No. STAN-CS-86-1106*, Stanford University, Stanford, CA, April 1986.

[Akiy83]    S.Akiyama, S.Ogawa, M.Yoneda, N.Yoshii, and Y.Terui, "Multilayer CMOS Device Fabricated on Laser Recrystallized Silicon Islands," *IEDM Tech. Dig.*, Dec. 1983.

[Barr84]    H.G.Barrow, "Verify: A Program for Proving Correctness of Digital Hardware Designs," *Artificial Intelligence*, Vol. 24, 1984.

[Bayo87]    M.A.Bayoumi, Class Discussions on VLSI Arrays, Univ. of Southwestern Louisiana, Lafayette, LA, 1987.

[Bayo89]    M.A.Bayoumi and Nam Ling, "Testing of a NORA CMOS Serial-Parallel Multiplier," *IEEE Journal of Solid-State Circuits*, Vol. 24, No. 2, pp. 494-503, April 1989.

[Bell71]    C.G.Bell and A.Newell, *Computer Structures: Readings and Examples*, McGraw-Hill, New York, 1971.

[Boch82]    G.V.Bochmann, "Hardware Specification with Temporal Logic: An Example," *IEEE Trans. on Computers*, Mar. 1982.

[Brow86]    M.C.Browne et al., "Automatic Verification of Sequential Circuits Using Temporal Logic," *IEEE Trans. on Computers*, Dec. 1986.

[Broc92]    B.C.Brock, W.A.Hunt,Jr., and W.D.Young, "Introduction to Formally Defined Hardware Description Language," *Theorem Provers in Circuit Design: Theory, Practice and Experience: Proceedings of the IFIP WG 10.2 International Conference*, Nijmegen, June 1992, ed. by V.Stavridon, T.F.Melham, and R.T.Boute, North Holland, 1992.

[Bush92]    H.Bush, "Transformational Design in Theorem Provers," *Theorem Provers in Circuit Design: Theory, Practice and Experience: Proceedings of the IFIP WG 10.2*

*International Conference*, Nijmegen, June 1992, ed. by V.Stavridon, T.F.Melham, and R.T.Boute, North Holland, 1992.

[Camu88]    P.Camurati and P. Prinetto, "Formal Verification of Hardware Correctness: Introduction and Survey of Current Research," *IEEE Computer Magazine*, July 1988.

[Capp83]    P.R.Cappello and K.Steiglitz, "Unifying VLSI Array Design with Geometric Transformations," *Proc. of the 1983 Int. Conf. on Parallel Processing*, Aug. 1983.

[Capp84]    P.R.Cappello and K.Steiglitz, "Unifying VLSI Array Design with Linear Transformations of Space-Time," *Advances in Computing Research*, Vol. 2, 1984.

[Chen82]    M.C.Chen and C.A.Mead, "Concurrent Algorithms as Space-Time Recursion Equations," *USC Workshop on VLSI and Modern Signal Processing*, 1982.

[Chen83]    M.C.Chen, *Space-Time Algorithms: Semantics and Methodology*, Ph.D. Thesis, California Institute of Technology, Pasedena, CA, May 1983.

[Dasg87]    S.Dasgupta, Class Discussions on Advanced Computer Architectures, Univ. of Southwestern Louisiana, Lafayette, LA, 1987.

[Dasg88a]   S.Dasgupta, Class Discussions on Methodology of Design, Univ. of Southwestern Louisiana, Lafayette, LA, 1988.

[Dasg88b]   S.Dasgupta, *Computer Architecture: A Modern Synthesis*, Vol. I & II, John Wiley & Sons, New York, 1988.

[Dule68]    J.R.Duley and D.L.Dietmeyer, "A Digital System Design Language (DDL)," *IEEE Trans. on Computers*, Vol. C-17, 1968.

[Etch81]    R. D. Etchelles et al., "Development of a Three-Dimensional Circuit Integration Technology and Computer Architecture," *Proc. SPIE*, Vol. 282, April 1981.

[Evek85]    H. Eveking, "The Application of CHDL's to the Abstract Specification of Hardware," *CHDL '85: IFIP 7th Int. Conf. on Comp. Hardware Description Lang. and their Applications*, Aug. 1985.

[Fost81]    M.J.Foster, "Syntax Directed Verification of Circuit Functions," *VLSI Systems and Computations*, ed. by H.T.Kung, B.Sproull, and G.Steele, Computer Science Press, Rockville, MD, 1981.

[Fuji83]    M.Fujita, H.Tanaka, and T.Moto-oka, "Verification with Prolog and Temporal Logic," *CHDL '83: IFIP 6th Int. Symp. on Comp. Hardware Description Lang. and their Applications*, May 1983.

[Gene87]    M.R.Genesereth and N.J.Nilsson, *Logical Foundations of Artificial Intelligence*, Morgan Kaufmann, Los Altos, CA, 1987.

[Gonc83]    N.F.Goncalves and H.J.D. Man, "NORA: A Racefree Dynamic CMOS Technique for Pipelined Logic Structures," *IEEE Journal of Solid-State Circuits*, Vol. SC-18, June 1983.

[Gord83]    M.J.C.Gordon, "LCF-LSM," *Tech. Report* No.41, Comp. Lab., Univ. of Cambridge, Cambridge, U.K., 1983.

[Gord86]    M.J.C.Gordon, "Why High-Order Logic Is a Good Formalism for Specifying and Verifying Hardware," *Formal Aspect VLSI Design: Proc. 1985 Edinburgh Conf. VLSI*, ed. by G.J.Milne and P.A.Subrahmanyam, North-Holland Pub., Amsterdam, Netherlands, 1986.

[Grin84]    J.Grinberg et al., "A Cellular VLSI Architecture for Image Analysis and Two-Dimensional Signal Processing," *IEEE Computer Mag.*, Jan. 1984.

[Gupt92]    A.Gupta, "Formal Hardware Verification Methods: A Survey," *Formal Methods in System Design*, Vol. 1, Nos. 2-3, Oct. 1992.

[Hail80]    B.T.Hailpern, "Verifying Concurrent Processes Using Temporal Logic," *Tech. Report No. 195*, Stanford University, Stanford, CA, 1980.

[Halp83]    J.Halpern et al., "A Hardware Semantics Based on Temporal Intervals," *Proc. of the Int. Coll. on Aut., Lang., and Prog.*, Spain, July 1983.

[Hann86]    F.K.Hanna and N.Daeche, "Specification and Verification of Digital Systems using Higher-order Logic," *IEE Proc.*, Vol. 133, Pt. E, No. 5, Sept. 1986.

[Henn86]    M.Hennessy, "Proving Systolic System Correct," *ACM Trans. on Prog. Lang. And Syst.*, July 1986.

[Hoare92]   C.A.R.Hoare and M.J.C.Gordon, eds., *Mechanized Reasoning and Hardware Design*, Prentice-Hall International Series in Computer Science, Prentice-Hall, 1992.

[Inou86]    Y.Inoue et al., "A Three-Dimensional Static RAM," *IEEE Electron Device Lett.*, Vol. EDL-7, May 1986.

[Jone63]    J.C.Jones, "A Method of Systematic Design," *Conference on Design Methods*, ed. by J.C.Jones and D.Thornley, Pergamon Press, Oxford, 1963.

[Jove84]    J.M.Jover and T.Kailath, "Design Framework for Systolic-Type Arrays," *Proc. of the 1984 IEEE Int. Conf. on Acoustics, Speech, and Signal Processing*, San Diego, Mar. 1984.

[Jove86]     J.M.Jover, T.Kailath, H.Lev-Ari, and S.K.Rao, "On the Analysis of Synchronous Computing Arrays," *1986 IEEE Workshop on VLSI Signal Processing II*, IEEE Press, 1986.

[Karp67]     R.P.Karp, R.E.Miller, and S.Winograd, "The Organization of Computation for Uniform Recurrence Equations," *Journal of the ACM*, Vol. 14(3), July 1967.

[Kell82]     E.Kelly and L.Steinberg, "The CRITTER System: Analyzing Digital Circuits by Propagating Behaviors and Specifications," *Proc. of the Nat. Conf. On Art. Int.*, AAAI-82(1982).

[KungH78]    H.T.Kung and C.E.Leiserson, "Systolic Arrays (for VLSI)," *Sparse Matrix Symp.*, SIAM, 1978.

[KungH80]    H.T.Kung, "Special Purpose Devices for Signal and Image Processing: An Opportunity in Very Large Scale Integration (VLSI)," *Proc. of the SPIE*, Vol. 241, Real-Time Signal Processing III, July 1980.

[KungH82]    H.T.Kung, "Why Systolic Architectures?" *IEEE Computer Mag.*, Jan. 1982.

[KungH83]    H.T.Kung and W.T.Lin, "An Algebra for VLSI Algorithm Design," *Proc. of the Conf. on Elliptic Problem Solvers*, Monterey, CA, Jan. 1983.

[KungS83]    S.Y.Kung, "From Transversal Filter to VLSI Wavefront Array," *Proc. of Int. Conf. on VLSI 1983*, IFIP, Trondheim, Norway, 1983.

[KungS84]    S.Y.Kung, "On Supercomputing with Systolic/Wavefront Array Processors," invited paper, *Proc. of the IEEE*, Vol. 72(7), July 1984.

[KungS85]    S.Y.Kung, "VLSI Array Processors," *IEEE ASSP Mag.*, July 1985.

[KungS86]    S.Y.Kung, P.S.Lewis, and S.C.Lo, "On Optimally Mapping Algorithms to Systolic Arrays with Applications to the Transitive Closure Problem," *Proc. of the 1986 IEEE Int. Symp. on Circuits and Systems*, 1986.

[KungS88]    S.Y.Kung, *VLSI Array Processors*, Prentice-Hall, Englewood Cliffs, New Jersey, 1988.

[Kuo84]      C.J.Kuo, B.C.Levy, and B.R.Musicus, "The Specification and Verification of Systolic Wave Algorithms," *1984 IEEE Workshop on VLSI Signal Processing I*, IEEE Press, 1984.

[Lam85]      M.Lam and J.Mostow, "A Transformational Model of VLSI Systolic Design," *IEEE Computer Mag.*, Feb. 1985.

[Leis83]    C.E.Leiserson, F.M.Rose, and J.B.Saxe, "Optimizing Synchronous Circuitry by Retiming," *Proc. of the 3rd Caltech Conf. on VLSI*, 1983.

[Lev83]    H.Lev-Ari, "Modular Computing Networks: A New Methodology for Analysis and Design of Parallel Algorithms/Architectures," *Integrated Systems Inc., Report No. 29*, Palo Alto, CA, Dec. 1983.

[Lind84]    R.W.Linderman and W.H.Ku, "A Three Dimensional Systolic Array Architecture for Fast Matrix Multiplication," *Proc. of the 1984 IEEE Int. Conf. on Acoustics, Speech and Signal Processing*, Vol. 2, 1984.

[Ling87]    Nam Ling and M.A.Bayoumi, "An Efficient Technique to Improve NORA CMOS Testing," *IEEE Trans. on Circuits and Systems*, Vol. CAS-34, No.12, pp. 1609-1611, Dec. 1987.

[Ling88]    Nam Ling and M.A.Bayoumi, "Algorithms for High Speed Multi-Dimensional Arithmetic and DSP Systolic Arrays," *Proc. of the 1988 Int. Conf. on Parallel Processing*, St. Charles, Illinois, pp. 367-374, Aug. 1988.

[Ling89a]    Nam Ling and M.A.Bayoumi, "Mapping Algorithms onto Multi-Dimensional Systolic Arrays," Chapter 10 of *Progress in Computer Aided VLSI Design, Vol.2: Techniques*, ed. by G.W.Zobrist, Ablex, Norwood, New Jersey, July 1989.

[Ling89b]    Nam Ling and M.A.Bayoumi, "STA: A Tool for Systolic Array Reasoning," *Proc. of the 1989 IEEE Int. Symp. on Circuits and Systems*, invited paper, Portland, Oregon, pp. 461-464, May 1989.

[Ling89c]    Nam Ling and M.A.Bayoumi, "Systematic Algorithm Mapping for Multi-dimensional Systolic Arrays," *Journal of Parallel and Distributed Computing*, Vol. 7, No. 2, Academic Press, pp. 368-382, Oct. 1989.

[Ling89d]    Nam Ling and M.A.Bayoumi, "The Design and Implementation of Multidimensional Systolic Arrays for DSP Applications," *Proc. of the 1989 IEEE Int. Conf. on Acoustics, Speech, and Signal Processing*, Glasgow, Scotland, U.K., pp. 1142-1145, May, 1989.

[Ling90]    Nam Ling and M.A.Bayoumi, "Systolic Temporal Arithmetic: A New Formalism for Specification and Verification of Systolic Arrays," *IEEE Trans. on Computer-Aided Design of Integrated Circuits and Systems*, Vol. 9, No. 8, pp. 804-820, August 1990.

[Ling93]    Nam Ling and T.Shih, "VSTA: A Prolog-Based Formal Verifier for Systolic Array Designs," *Proceedings of the 1993 International Conference on Parallel Processing (ICPP)*, St. Charles, Illinois, USA, pp. II-73 to II-76, August 16-20, 1993.

[Ling95]      Nam Ling, "A Special Purpose Formal Verifier for Systolic Designs in DSP Applications," *Journal of VLSI Signal Processing*, Vol. 11, Nos. 1/2, Kluwer Academic Publishers, pp. 169-187, October/November 1995.

[Mala81]      Y.Malachi and S.S.Owicki, "Temporal Specifications of Self-Time Systems," *VLSI Systems and Computations*, ed. by H.T.Kung et al., Computer Science Press, 1981.

[Mann81]     Z.Manna and A.Pnueli, "Verification of Concurrent Programs: The Temporal Framework," *The Correctness Problem in Computer Science*, ed. by R.S.Boyer and J.S.Moore, Academic Press, 1981.

[Marc86]     M.Marcotty and H.F.Ledgard, *Programming Language Landscape*, Science Research Associates, Inc., 1986.

[Mead80]     C.Mead and L.Conway, *Introduction to VLSI Systems*, Addison-Wesley, Reading, MA, 1980.

[Melh84]     R.G.Melhem and W.C.Rheinboldt, "A Mathematical Model for the Verification of Systolic Networks," *SIAM J. of Comput.*, Vol. 13, No. 3, Aug. 1984.

[MelT88]     T.F.Melham, "Abstraction Mechanisms for Hardware Verification," *VLSI Specification, Verification and Synthesis*, ed. by G.M.Birtwistle and P.A.Subrahmanyam, Kluwer Academic Publishers, 1988.

[MelT93]     T.F.Melham, *Higher Order Logic and Hardware Verification*, Cambridge University Press, 1993.

[Mil83]       G.J.Milne, "Circal: A Calculus for Circuit Description," *Integration, the VLSI Journal*, July 1983.

[Miln80]      R.Milner, "A Calculus of Communicating Systems," *Lecture Notes in Computer Science, 92*, Springer-Verlag, New York, 1980.

[Miln83]      R.Milner, "Calculi for Synchrony and Asynchrony," *Theor. Comput. Sci. 25*, 1983.

[Mira84]      W.L.Miranker, "Space-time Representation of Computational Structures," *Computing*, 1984.

[Mits85]      K.Mitsuhashi, "Etch Back Planarization Technique and Its Application to Multilayer Devices," *Proc. 4th FED Symp.*, July 1985.

[Mold83]     D.I.Moldovan, "On the Design of Algorithms for VLSI Systolic Arrays," *Proc. of the IEEE*, Vol. 71, No. 1, Jan. 1983.

[Mold84]     D.I.Moldovan, "ADVIS: A Software Package for the Design of Systolic Arrays," *Proc. of the IEEE Int. Conf. on Computer Design*, 1984.

[Mold86]    D.I.Moldovan and J.A.B.Fortes, "Partitioning and Mapping of Algorithms into Fixed Size Systolic Arrays," *IEEE Trans. on Computers*, Vol. 35(1), Jan. 1986.

[Mosz83]    B.Moszkowski and Z.Manna, "Reasoning in Interval Temporal Logic," *Report No. STAN-CS-83-969*, Stanford University, Stanford, CA, July 1983.

[Mosz85]    B.Moszkowski, "A Temporal Logic for Multilevel Reasoning about Hardware," *IEEE Computer Mag.*, Feb. 1985.

[Nudd85]    G.R.Nudd et al., "Three-Dimensional VLSI Architecture for Image Understanding," *Journal of Parallel and Distributed Computing*, Vol.2, 1985.

[Osse82]    M.Ossefort, "Correctness Proofs of Communicating Processes-Three Illustrative Examples from the Literature," *TR-LCS-8201*, Dept. of Computer Science, Univ. of Texas, Austin, TX, Jan. 1982.

[Pann85]    G.Panneerselvam, "Three Dimensional Systolic Cubic Architecture for Simultaneous Triple Matrix Multiplication," *Proc. of 1st Int. Conf. On Supercomputing Systems*, Dec. 1985.

[Prep81]    R.P.Preparata and J.Vuillemin, "The Cube-Connected Cycles: A Versatile Network for Parallel Computation," *Comm. of ACM*, Vol. 24, May 1981.

[Prob88]    D.K.Probst and H.F.Li, "Abstract Specification of Synchronous Data Types for VLSI and Proving the Correctness of Systolic Network Implementations," *IEEE Trans. on Computers*, Vol. 37, No. 6, June 1988.

[Puru89]    S.Purushothaman and P.A.Subrahmanyan, "Mechanical Certification of Systolic Algorithms," *Journal of Automated Reasoning*, Kluwer Academic Publishers, Mar. 1989.

[Quin84]    P.Quinton, "Automatic Synthesis of Systolic Arrays from Uniform Recurrent Equations," *Proc. of the 11th Annual Symp. on Computer Architecture*, 1984.

[Rajo85]    S.V.Rajopadhye and P.Panagaden, "Verification of Systolic Arrays: A Stream Functional Approach," *UUCS-85-001*, University of Utah, Salt Lake City, UT, Mar. 1985.

[Rao85]     S.K.Rao, *Regular Iterative Algorithms and Their Implementation on Processor Arrays*. Ph.D. Dissertation, Stanford University, Stanford, CA, 1985.

[Res83]     N.Rescher and A.Urquart, *Temporal Logic*, Springer-Verlag, New York 1971.

[Rose83]    A.L.Rosenberg, "Three-Dimensional VLSI: A Case Study," *Journal of the ACM*, Vol. 30, July 1983.

[Shah85]    M.Shahdad et al., "VHSIC Hardware Description Language," *IEEE Computer Mag.*, Feb. 1985.

[Shih95]    T.Shih, Nam Ling, R.Davis, and F.Lin, "On the Construction of a Prolog-Based Verifier for Systolic Array Designs," *Computational Intelligence: An International Journal*, Vol. 11, No. 1, Blackwell Publishers, pp. 172-221, February 1995.

[Shos83]    R.E.Shostak, "Formal Verification of Circuit Designs," *CHDL'83: IFIP 6th Int. Symp. on Comp. Hardware Description Lang. and their Applications*, May 1983.

[Stau94]    J.Staunstrup, *A Formal Approach to Hardware Design*, Kluwer Academic Publishers, 1994.

[Tera87]    A.Terao and F.V.d.Wiele, "Purposes of Three-Dimensional Circuits," *IEEE Circuits and Devices Mag.*, Nov. 1987.

[Tid84]     E.Tiden, "Verification of Systolic Arrays-A Case Study," *Technical Report TRITA-NA-8403*, Dept. of Numerical Analysis and Computer Science, The Royal Institute of Technology (Sweden), 1984.

[Turn84]    R.Turner, *Logics for Artificial Intelligence*, Ellis Horwood Limited, Chichester, West Sussex, England, U.K., 1984.

[Ueha83]    T.Uehara et al., "DDL Verifier and Temporal Logic," *CHDL'83: IFIP 6th Int. Symp. on Comp. Hardware Description Lang. and their Applications*, May 1983.

[Ull84]     J.D.Ullman, *Computational Aspects of VLSI*, Computer Science Press, 1984.

[Wag77]     T.Wagner, "Hardware Verification," *Tech. Report STAN-CS-77-632*, Stanford University, Stanford, CA, Sept. 1977.

[Wilse87]   P.A.Wilsey et al., "An S*M Execution Environment," *Tech. Report TR87-3-1*, Univ. of Southwestern Louisiana, Lafayette, LA, 1987.

[Wilso83]   P.Wilson, "Occam Architecture Eases System Design," *Computer Design*, Nov. 1983.

[Woj84]     A.S.Wojcik, "A Formal Design Verification System Based on an Automated Reasoning System," *Proc. of the ACM IEEE 21st Design Automation Conf.*, July 1984.

[Wolp81]    P.Wolper, "Temporal Logic can be more Expressive," *Proc. of the 22nd Annual Symp. on Foundation of Computer Science*, October 1981.

[Yoe90]     M. Yoeli ed., *Formal Verification of Hardware Design*, IEEE Computer Society Press, 1990.

# AUTHOR BIOGRAPHIES

**Nam Ling** received a B.Eng degree from the National University of Singapore. He received an M.S. degree and a Ph.D. degree, both in computer engineering, from the Center for Advanced Computer Studies at the University of Southwestern Louisiana, Lafayette, Louisiana, U.S.A. After completing his Ph.D. in 1989, Dr. Ling joined the Computer Engineering Department, Santa Clara University, California, U.S.A, where he is currently an Associate Professor. He was a Senior Research Fellow, visiting the Center for Signal Processing, Nanyang Technological University, in Singapore, during 1998.

Dr. Ling has published more than 70 research papers in major refereed journals, conferences, and books in the fields of image and video compression, systolic arrays, as well as parallel processing for DSP. Dr. Ling received research support and funding from U.S. National Science Foundation and the industry in Silicon Valley. He received Arthur Vining Davis Junior Faculty Fellowship in 1991 and the Santa Clara University Outstanding Achievement Award in Teaching, Research, and Service in 1992.

Besides research, Dr. Ling teaches courses in multimedia data compression, parallel processing, architecture, and logic design. He is among the Who's Who Among America's Teachers. In 1993-1995, Dr. Ling served as the Chair of the IEEE Computer Society Technical Committee on Microprocessors and Microcomputers. Dr. Ling was the General Chair of the IEEE Hot Chips Symposium in 1995, drawing an attendance of about 1,000. He received the IEEE Computer Society Certificate of Appreciation in 1997. He also participated in the Singapore Digital TV Technical Committee in 1998. Dr. Ling served in program committees and as session chairs for several conferences. Dr. Ling is an IEEE Senior Member.

**Magdy A. Bayoumi** is an Edmiston Professor of Computer Engineering in the Center for Advanced Computer Studies, at the University of Southwestern Louisiana (USL), where he has been a faculty member since 1985. Dr. Bayoumi received the B.Sc. and M.Sc. degrees in Electrical Engineering from Cairo University, Egypt; M.Sc. degree in Computer Engineering from Washington University, St. Louis; and the Ph.D. degree in Electrical Engineering from the University of Windsor, Canada. Dr. Bayoumi's research interests include VLSI Design Methods and Architectures, Low Power Circuits and Systems, Digital Signal Processing Architectures, Parallel Algorithm Design, Computer Arithmetic, Image and Video Signal Processing, Neural Networks and Wideband Network Architectures.

Dr. Bayoumi is the Vice President for the Technical Activities of the IEEE Circuits and Systems Society. He is a founding member of the VLSI Systems and Applications Technical Committee and was its Chairman. He is a member of the Neural Network and the Multimedia Technology Technical Committees. He has been on the technical program committee for ISCAS for several years, and he is the Publication Chair for ISCAS'99. He was the General Chairman for 1994 MWSCAS and he is a member of the Steering Committee of this symposium. He was an Associate Editor of the IEEE Circuits and Devices Magazine, the IEEE Transactions on VLSI Systems, and the IEEE Transactions on Neural Networks. He is an Associate Editor of the IEEE Transactions on Circuits and Systems II.

Dr. Bayoumi serves on the ASSP Technical Committee on VLSI Signal Processing. He was the Co-Chairman of the Workshop on Computer Architecture for Machine Perception, 1993 and he is a member of the Steering Committee of this workshop. He was the General Chairman for the 8th Great Lake Symposium on VLSI, 1998. Dr. Bayoumi is an Associate Editor of INTEGRATION, the VLSI Journal, and the Journal of VLSI Signal Processing. He is a regional editor for the VLSI Design Journal and he is on the Advisory Board of the Journal on Microelectronics Systems Integration. Dr. Bayoumi has edited and co-edited three books in the area of VLSI Signal Processing. Dr. Bayoumi served on the Distinguished Visitors Program for the IEEE Computer Society, 1991-1994. He is the faculty advisor for the IEEE Computer student chapter at USL. He won the USL 1988 Researcher of the Year award and the 1993 Distinguished Professor award at USL. He is an IEEE fellow.

# INDEX